VSWR and Impedance Matching Techniques

Ain Rehman
Signal Processing Group Inc.

Preface

In the design of radio frequency or wireless circuits the concept of power and power transfer is key. Radio and high frequency parameters are usually based on power quantities unlike that in the low frequency regime where voltages and currents are usually deemed appropriate and sufficient.

Power transfer from the source to the load is so important that a number of parameters and critical concepts have evolved to analyze the consequences of this important factor. Among these are the Standing Wave Ratio (Voltage Standing Wave Ratio, Current Standing Wave Ratio and Power Standing Wave Ratio). Reflection Coefficients, Return Loss, Mismatch loss etc. All are related to each other and are in frequent use in the high frequency design community.

The understanding of these quantities and their interplay is an essential part of both the practicing engineer and the student's repertoire. Without a solid grounding in these principles the practitioner's art becomes difficult to say the least.

This eCADbook^{TM} is another effort to present these important concepts. The difference is, that this book chooses the path of a concise approach where extended mathematical derivations are downplayed and the concepts and parameters are presented in a cookbook fashion. However, mathematics is not completely ignored, but is embodied in a number of CAD routines and scripts (Javascript, C++/Windows) which the reader is encouraged to use in the reading of the book and in subsequent work. In this way the "wood is not obscured by the trees" and the reader can try examples from the book or those in his/her own work to allow a deeper understanding of the principles. Also time is saved by the use of these software tools.

Part one deals with brief intuitive descriptions of the quantities in question. This work lays the foundation for part two which enunciates the mathematical expressions and formulas for these parameters. Part three addresses the lumped element impedance matching approach to power transfer. The concept of impedance matching is very important since high frequency design sometimes becomes the art of matching impedances for proper operation of the circuit blocks and antennas. Part four continues the impedance matching discussion using the concept of the Smith Chart and transmission lines. In this case the matching elements can be deemed distributed. There is significant discussion of discrete lumped elements and transmission line quantities including microstrip lines.

For each topic under discussion, there are a number of CAD routines and scripts included. Finally it should be understood that _not every impedance matching technique_ is addressed in this book. Only the more popular ones. However the assumption is that once one becomes familiar with the techniques presented here further reading and research may become easier to understand, both intuitively and in terms of mathematical approaches.

CONTENTS

_____

Part I

The Basics

A signal propagates on wires or transmission lines that connect a source of power to the load which uses this power to do useful work. In so doing a number of electronic phenomena occur, such as reflection of the input signal that causes a loss of useful power which sometimes dissipates harmlessly or in some critical cases disastrously by generating reflective effects that can destroy the circuit or the equipment it is included in.

To understand and quantify these concepts a number of measures have been developed over time that are in common use throughout the design, standards and manufacturing community, so that a common framework exists, and is accepted by, and understood by, the majority of practitioners in the industry. These basic concepts are described below.

1.1   Propagation on transmission lines

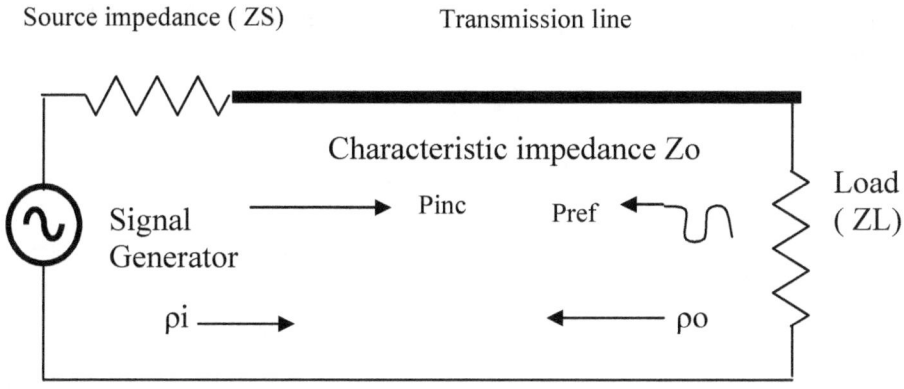

Figure B.1.0

This section presents a semi-intuitive view of the quantities of signal propagation on transmission lines with just enough mathematics to illustrate the concepts involved.

A transmission line can be assumed to have an equivalent circuit involving R, L,C and G elementary parts (See figure B2.0). In addition, a transmission line also has the delay attribute where signals traveling on the line get delayed.

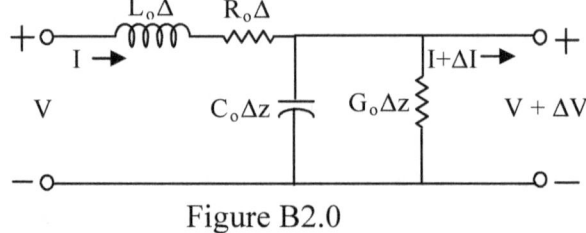

Figure B2.0

Lo, Ro, Co and Go are constants of the line and are evaluated per unit length. The total values of these elements are then the product of the length of the line and the elementary constants.

V and I are the voltage and current traveling on the line. The incremental voltage and current is also shown in Figure B2.0. These may be deemed to be intermediate values and again can be evaluated over the total length of the line.

1.2 Standing waves, the reflection coefficient and the Standing Wave Ratio.

In addition to this, if the terminations on the line (ZS, ZL) are not _exactly conjugately matched,_ reflection of the incident signal (Pinc) occurs at the load end, and a reflected signal (Pref) travels back towards the source. This can set up standing waves on the line as shown below. ρi and ρo are the reflection coefficient magnitudes. The reflection coefficient is defined as the _ratio of the reflected power by the incident power._ ρi is the reflection coefficient magnitude for the reflection of source power with a mismatched transmission line characteristic impedance and ρo is the load end reflection coefficient that represents the mismatch with the load and the characteristic impedance of the line.

The characteristic impedance Zo is the uniform impedance of the line that depends on the line's physical constants and the dielectric medium in which the line is resident. In case of microstrip lines the characteristic impedance depends on the width, height and effective dielectric constant of the microstrip line. More detail on this is presented later on.

An understanding of the reflection coefficient and the VSWR can be facilitated by using the following equation (presented without proof) and using a simple program to generate a graphical look at standing waves.

$$V(t, x) = A\sqrt{4\rho\cos^2 \beta x + (1 - \rho)^2}\ \cos(\omega t + \theta), \quad (B1.0)$$

where ρ is the reflection coefficient magnitude, β is the wave number (angular frequency/wave phase velocity; alternatively $2\pi/\lambda$, where λ is the wavelength) and θ is the phase and is given by $\dfrac{(1+\rho)}{(1-\rho)}\cot(\beta x)$. As can be seen the voltage is both a function of x, the distance and the time.

The voltage variation at any point on a transmission line with arbitrary loads is:

$$V = A\sqrt{4\rho\cos^2 \beta x + (1-\rho)^2} \qquad \text{(B2.0)}$$

where A is the signal amplitude. This equation can be implemented in a simple computer script and the results can be plotted to give a glimpse into the behavior of standing waves. A true plot would include both the time variation, as well as variation with distance. However, a plot of the standing waves with respect to distance is good enough to further the understanding of standing waves and the VSWR.

In the plots shown we have set $\beta = 1$ and the amplitude $A = 1$ without loss of generality since our intention is to look at the effect of ρ on the standing waves. Figure B3.0 shows the case for a reflection coefficient of 1.0.

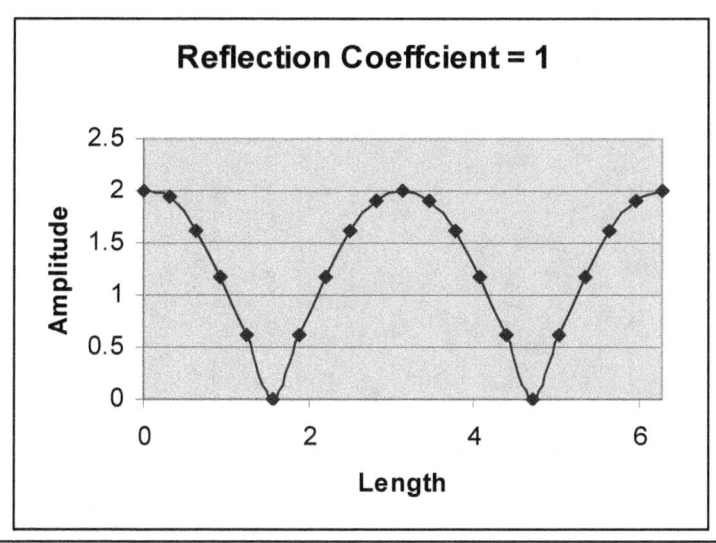

$\rho = 1$. Complete reflection, SWR = infinite : Figure B3.0

Figure B4.0 shows the case when the incident signal is completely dissipated by the load. In this case there is no reflected signal and the VSWR is 1.0, i.e, a perfectly matched case.

Figure B5.0 shows the case of $\rho = 0.5$, while figure B6.0 shows the case for $\rho = 0.2$. From these figures and analysis, a graphical understanding of the effect of the reflection coefficient can be developed. The reader may run his own plots with different values of the reflection coefficient using the GNUPLOT scripts as shown. Just type in the script using the plot command <script> and make sure that the sampling rate is at least 1000.

Using this technique, the effects of reflection can be analyzed in detail. The SWR is simply the maximum value of the amplitude divided by the minimum value of the amplitude.

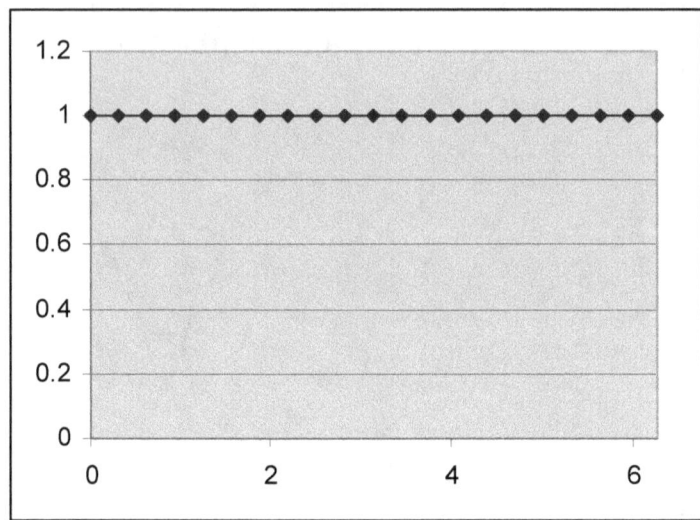

p = 0. Complete absorption, SWR = 1: Figure B4.0

The plots are generated with the length of line in radians along the
x – axis, and the amplitude along the y – axis. In these plots the incident
voltage amplitude is set to unity and the wave number is also set to unity for
clarity.

Generally the measurements of this type are done using the slotted line
technique. The slotted line is simply a transmission line containing a
movable probe. The movable probe can be set at various positions along the
line. The probe measures the voltages at these positions. Knowing the
voltage maxima and minima with length, a number of important parameters
can be found, such as the reflection coefficient, VSWR, value of the
unknown load impedance and so on.

The characteristics of the current along the line can also be found. It should
be obvious that, where on the length of line, the traveling waves add to
generate the voltage maxima, they also subtract to generate the current
minima. The maximum voltage position is also the minimum current
position. The impedance at the current minima is at the maximum
impedance. This maximum impedance can be written as:

$$\underline{ZMAX = Zo.VSWR.}$$

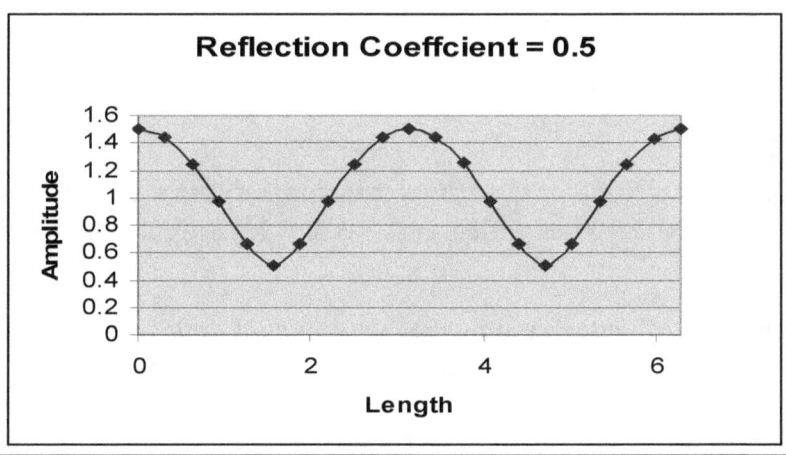

$\rho = 0.5$. Partial reflection , SWR = 3.0 : Figure B5.0

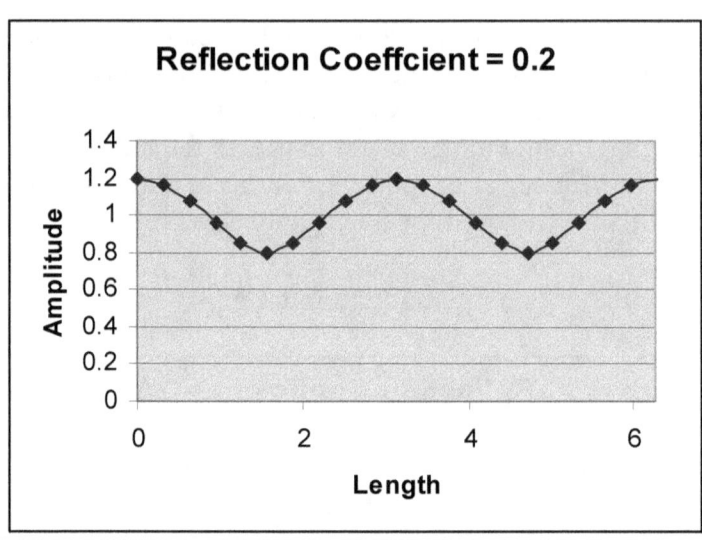

ρ = 0.2. Partial reflection , SWR = 1.5 : Figure B6.0

With this level of understanding we can move on to the next part of the eCADbook™ and look at various other quantities and parameters involved.

Part II
Parameters and expressions

VSWR, Reflection coefficient, Return loss, s11/s22.

2.1   VSWR:

The SWR is usually defined as a voltage ratio called the **VSWR**, for *voltage standing wave ratio*. For example, the VSWR value 1.2:1 denotes a maximum standing wave amplitude that is 1.2 times greater than the minimum standing wave value. It is also possible to define the SWR in terms of current, resulting in the ISWR, which has the same numerical value. The *power standing wave ratio* (PSWR) is defined as the square of the VSWR.

The VSWR is related to the reflection coefficient as:

$$\text{VSWR} = \frac{V_{max}}{V_{min}} = \frac{1+\rho}{1-\rho}$$

where ρ = the magnitude of the reflection coefficient.

2.2   Reflection coefficient:

Reflections occur as a result of discontinuities, such as an imperfection in an otherwise uniform transmission line, or when a transmission line is terminated with other than its characteristic impedance. The reflection coefficient Γ is defined thus: (V_r = reflected voltage, V_f = incident voltage)

$$\Gamma = \frac{V_r}{V_f}$$

Γ is a complex number that describes both the magnitude and the phase shift of the reflection. The simplest cases, when the imaginary part of Γ is zero, are:

- $\Gamma = -1$: maximum negative reflection, when the line is short-circuited.
- $\Gamma = 0$: no reflection, when the line is perfectly matched.
- $\Gamma = +1$: maximum positive reflection, when the line is open-circuited.

For the calculation of VSWR, only the magnitude of Γ, denoted by ρ, is of interest. Therefore, we define:

$$\rho = |\Gamma|.$$

2.3   Return loss:

Return loss or **Reflection loss** is the reflection of signal power resulting from the insertion of a device in a transmission line or optical fiber. It is usually expressed as a ratio in dB relative to the transmitted signal power.

If the power transmitted by the source is P_T and the power reflected is P_R, then the return loss in dB is given by:

$$RL(dB) = 10\log_{10}\frac{P_T}{P_R}$$

Optical Return Loss is a positive number, historically ORL has also been referred to as a negative number. Within the industry expect to see ORL referred to variably as a positive or negative number.

This ORL sign ambiguity can lead to confusion when referring to a circuit as having high or low return loss; so remember:- High Return Loss = lower reflected power = large ORL number = generally good. Low Return Loss = higher reflected power = small ORL number = generally bad.

In metallic conductor systems, reflections of a signal traveling down a conductor can occur at a discontinuity or impedance mismatch. The ratio of the amplitude of the reflected wave V_r to the amplitude of the incident wave V_i is known as the reflection coefficient Γ.

$$\Gamma = \frac{V_r}{V_i}$$

When the source and load impedances are known values, the reflection coefficient is given by:

$$\Gamma = \frac{Z_L - Z_S}{Z_L + Z_S}$$

where Z_S is the impedance toward the <u>source</u> and Z_L is the impedance toward the <u>load.</u>

Return loss is simply <u>the magnitude of the reflection coefficient in dB</u>. Since power is proportional to the square of the voltage, then <u>return loss</u> is given by:

$$RL(dB) = -20 Log(|\Gamma|)$$

where the vertical bars indicate magnitude. Thus, <u>a large positive return loss indicates the reflected power is small relative to the incident power, which indicates good impedance match from source to load.</u>

When the actual transmitted (incident) power and the reflected power are known (i.e. through measurements and/or calculations), then the return loss in dB can be calculated as the difference between the incident power P_i (in dBm) and the reflected power P_r (in dBm).

$$RL(dB) = P_i(dBm) - P_r(dBm)$$

s11/s22 relationship to impedance matching:

2.4   Input return loss

Input return loss (RL_{in}) is a scalar measure of how close the actual input impedance of the network is to the nominal system impedance value and, expressed in logarithmic magnitude, is given by:

$$RL = |20Log_{10}(|s11|)| \text{ dB.}$$

By definition, return loss is a positive scalar quantity implying the 2 pairs of magnitude (|) symbols. The linear part, $|s11|$ is equivalent to the reflected voltage magnitude divided by the incident voltage magnitude.

2.5   Output return loss

The output return loss (RL_{out}) has a similar definition to the input return loss but applies to the output port (port 2) instead of the input port. It is given by:

$$RL_{out} = |20Log_{10}(|s22|)| \text{dB.}$$

2.6   Voltage reflection coefficient

The voltage reflection coefficient at the input port (ρ_{in}) or, at the output port (ρ_{out}) are equivalent to s11 and s22 respectively, so:

$$\rho_{in} = s11 \text{ and } \rho_{out} = s22$$

As s11 and s22 are complex quantities, so are ρ_{in} and ρ_{out}.

Voltage reflection coefficients are complex quantities and may be graphically represented by polar diagrams on Smith Charts.

2.7   Voltage standing wave ratio

The voltage standing wave ratio (VSWR) at a port, represented by the lower case 's', is a similar measure of port match to return loss but is a scalar linear quantity, the ratio of the standing wave maximum voltage to the standing wave minimum voltage. It therefore relates to the magnitude of the voltage reflection coefficient and hence to the magnitude of either s11 for the input port or s22 for the output port.

At the input port, the VSWR (Sin) is given by:

$$\text{Sin} = \frac{1 + |s11|}{1 - |s11|}$$

At the output port, the VSWR (Sout) is given by:

$$\text{Sout} = \frac{1 + |s22|}{1 - |s22|}$$

2.8   <u>**Relationship of VSWR to Return Loss:**</u>

VSWR and return loss are related quantities. Note that the reflection coefficient Γ can be written in terms of the VWSR as:

$$\Gamma = \frac{VSWR - 1}{VSWR + 1}$$

This expression can be derived as follows:

From the given relationship,

$$VSWR = \frac{V_{max}}{V_{min}} = \frac{1 + \rho}{1 - \rho}$$

We can show that ρ is given by:

$$\rho = |\Gamma|,$$

thus,

$$VSWR = \frac{1 + \Gamma}{1 - \Gamma}$$

Which provides the relationship we started with.

The return loss can then be cast in terms of VSWR as:

$$RL(DB) = -20\log_{10}\left[\frac{VSWR - 1}{VSWR + 1}\right]$$

Conversely

$$VSWR = \frac{\left[10^{\frac{RL(dB)}{20.0}}\right] + 1.0}{\left[10^{\frac{RL(dB)}{20.0}}\right] - 1.0}$$

So if either of the two quantities is known the other can be calculated from it.

Table 1.0 shows these conversions below for convenience.

This table was generated by using the conversion equations shown above.

Table 1.0

Return loss (db)	VSWR	Reflection coefficient
0	Infinite	1.0
1	17.39	0.891
2	8.724	0.794
3	5.848	0.707
4	4.419	0.630
5	3.569	0.562
6	3.009	0.501
7	2.614	0.446
8	2.322	0.398

9	2.099	0.354
10	1.924	0.316
11	1.784	0.281
12	1.670	0.251
13	1.576	0.223
14	1.498	0.199
15	1.432	0.177
16	1.376	0.158
17	1.328	0.141
18	1.288	0.125
19	1.252	0.112
20	1.222	0.100
20.8	1.195	0.089
21.7	1.179	0.082
22.6	1.16	0.074
23.1	1.15	0.069
23.7	1.139	0.065
24.3	1.129	0.060
24.9	1.120	0.056
25.7	1.109	0.051
26.4	1.100	0.047
27.3	1.109	0.043
28.3	1.079	0.038
29.4	1.07	0.033

30.7	1.06	0.029
32.3	1.049	0.024
34.1	1.04	0.019
36.6	1.03	0.014
40.1	1.019	0.009
46.1	1.009	0.004

It is instructive to examine a graphical view of the relationship between VSWR and the return loss presented below.

Note the slow variation of the return loss as the VSWR reaches between 1.0 to 2.0. Conversely for VSWR of 7 or 8 the return loss is low.

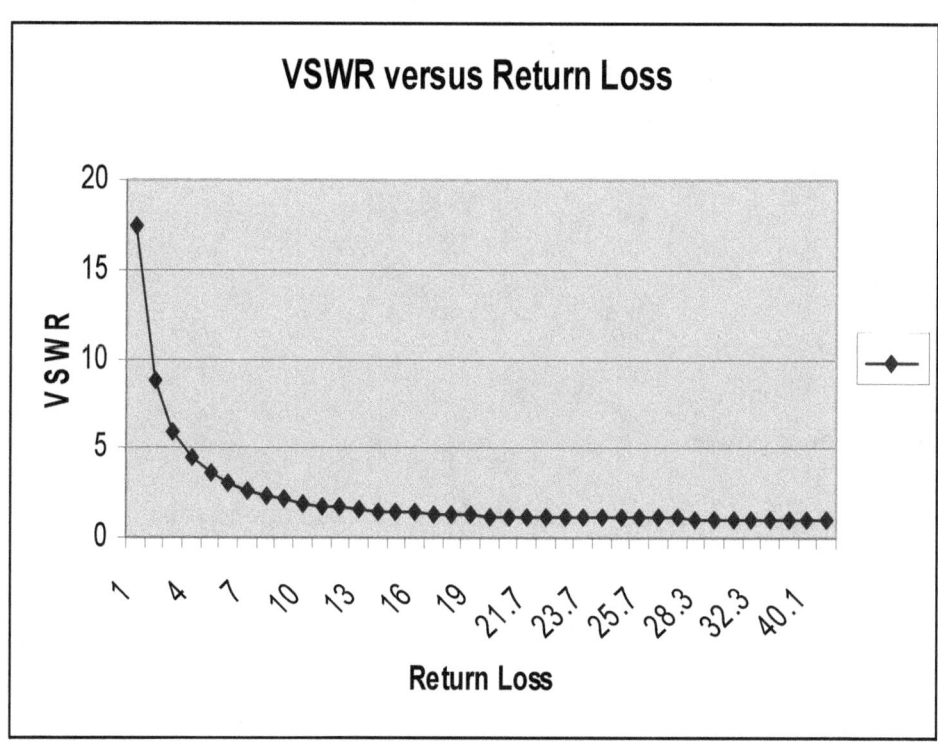

2.9   Mismatch loss:

Mismatch loss in transmission line theory is the amount of power expressed in decibels that will **not be available** at the output due to impedance mismatches and reflections.

Mismatch loss (ML) is the ratio of incident power to the difference between incident and reflected power:

$$\text{ML}_{dB} = 10\text{Log}_{10}\left(\frac{P_i - P_r}{P_i}\right)$$

$$P_r = P_i - P_d$$

where

P_i = incident power
P_r = reflected power
P_d = delivered power

The amount of incident power not reaching the load due to mismatching is:

$$\frac{P_d}{P_i} = 1 - \rho^2$$

where ρ is the reflection coefficient.

If the reflection coefficient is known, mismatch can be calculated by:

$$\text{ML}_{dB} = 10\text{Log}_{10}(1 - \rho^2)$$

In terms of the voltage standing wave ratio (VSWR):

$$ML_{dB} = 10Log_{10}\left(1.0 - \left(\frac{VSWR - 1}{VSWR + 1}\right)^2\right)$$

It is obvious from these expressions that mismatch loss can be calculated from VSWR and the reflection coefficient and vice versa.

Part III
<u>Lumped element impedance matching techniques</u>

3.1   Consider Figure 1.0 below. It represents the impedance
        matching problem.

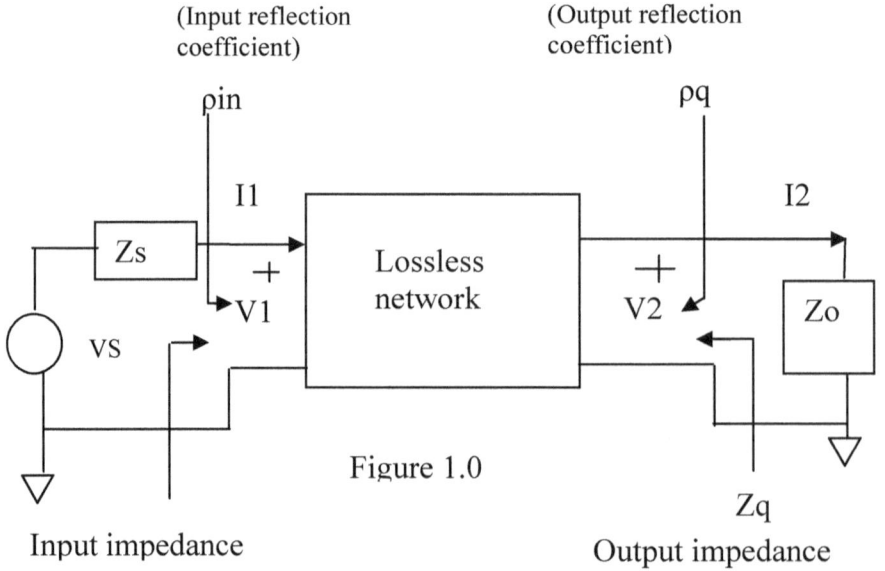

Figure 1.0

Input impedance   Output impedance

*<u>The art of impedance matching lies in the design of a network so that a
terminating impedance is transformed exactly to a desired impedance at a
frequency, or is transformed approximately over a band of frequencies.</u>*

An important concept is that:

   (a) $Z1 = Zs*$

                                           (1)

   and

   (b) $Z2 = Zq*$ (2)

Also the load impedance must be the complex conjugate of the source .

3.2 Narrow band pi, L and T networks.

The simplest matching networks are the pi, L and T networks shown below. The circuits are composed of reactance's as shown. R1 is the input resistance and R2 the output resistance.

Type A

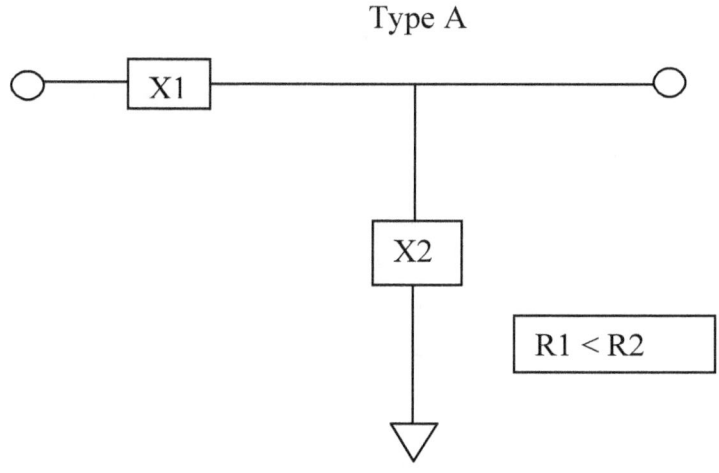

Type B

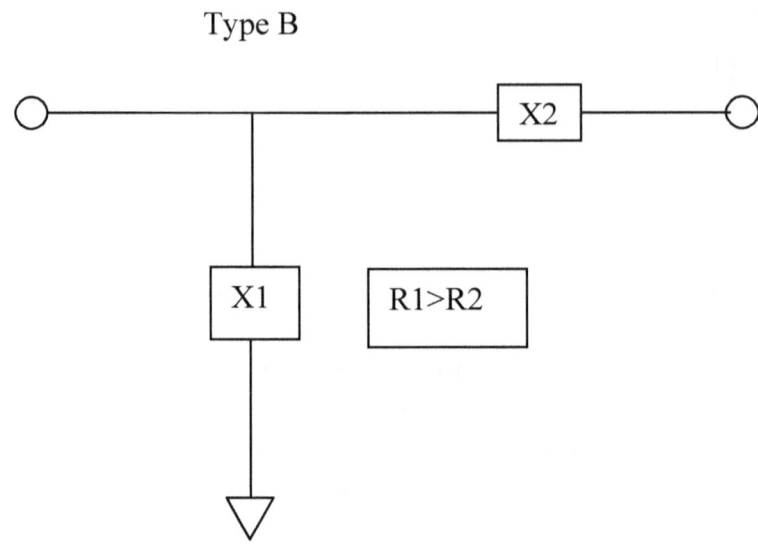

Type T

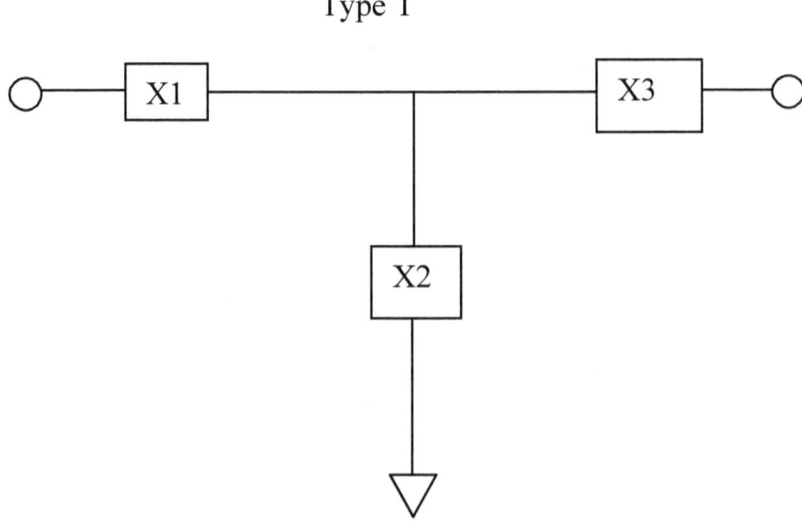

Type Pi

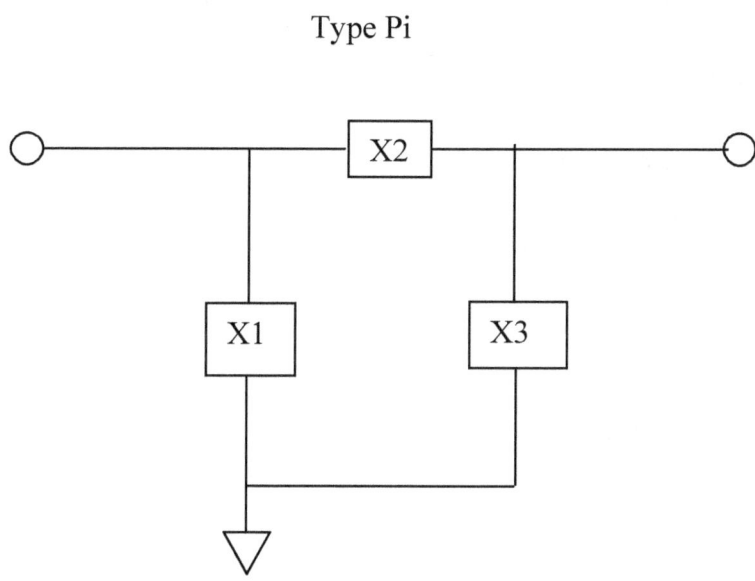

The **transfer phase** is defined as the angle by which the current I2 **lags** the current I1. If Z1 and Z2 are **resistive** the phase is same as the phase lag of V2 with respect to V1. The phase angles of type A and type B sections are **dependent** , assuming that R1 and R2 are **independent**. The phase is also independent for the T and Pi sections within a range of + 180 Degrees to - 180 Degrees (except zero). The design equations for these networks are shown below. Also, note that since the phase angle for the L sections is not independent it **cannot be chosen.** It will be what it will be, depending on R1 and R2. On the other hand the phase angle for the T and Pi networks **can be chosen** (or specified) as mentioned above.

The relationships for a *real source and real load* are shown below for the T, Pi and L networks.

T network:

$$X1 = \frac{\sqrt{R1R2} - R1\cos\beta}{\sin\beta} \qquad (3)$$

$$X2 = \frac{-\sqrt{R1R2}}{\sin\beta} \qquad (4)$$

$$X3 = \frac{\sqrt{R1R2} - R2\cos\beta}{\sin\beta} \qquad (5)$$

Pi network:

$$X1 = \frac{R1R2\sin\beta}{R2\cos\beta - \sqrt{R1R2}} \qquad (6)$$

$$X2 = \frac{R1R2\sin\beta}{\sqrt{R1R2}} \qquad (7)$$

$$X3 = \frac{R1R2\sin\beta}{R1\cos\beta - \sqrt{R1R2}} \qquad (8)$$

Note: In these networks the phase angle β can be chosen from 180 degrees to -180 degrees except 0 degrees.

Type A network:

$$X1 = \frac{\sqrt{R1R2} - R1\cos\beta}{\sin\beta} \qquad (9)$$

$$X2 = \frac{-\sqrt{R1R2}}{\sin\beta} \qquad (10)$$

Also the phase angle is defined by:

$$\beta = \pm\cos^{-1}\sqrt{\frac{R1}{R2}} = \tan^{-1}\sqrt{\frac{R2}{R1} - 1.0} \qquad (11)$$

Type B network:

$$X1 = \frac{R1R2\sin\beta}{R2\cos\beta - \sqrt{R1R2}} \qquad (12)$$

$$X2 = \frac{R1R2\sin\beta}{\sqrt{R1R2}} \qquad (13)$$

$$X3 = \infty. \qquad (14)$$

The phase angle is defined by:

$$\beta = \pm\cos^{-1}\sqrt{\frac{R2}{R1}} = \tan^{-1}\sqrt{\frac{R1}{R2} - 1.0} \qquad (14.1)$$

3.3   The Q matching technique.

Another approach to L section matching is the Q matching technique. This is explained below:

Two resistive terminations, one at the input to a network, and the other at its output can be simultaneously matched by adding two reactive elements between them. Calling the terminations Rlow and Rhigh the following can be done.

1)   **Add a series element next to Rlow and a parallel one next to Rhigh.**
2)   **The series element can be a capacitor or an inductor.**
3)   **The parallel element has to be of the opposite type.**
4)   **A series inductor with a parallel capacitor is a low pass circuit**
5)   **A series capacitor with a parallel inductor is a high pass circuit.**

Figures 1 through 4 below show the configurations

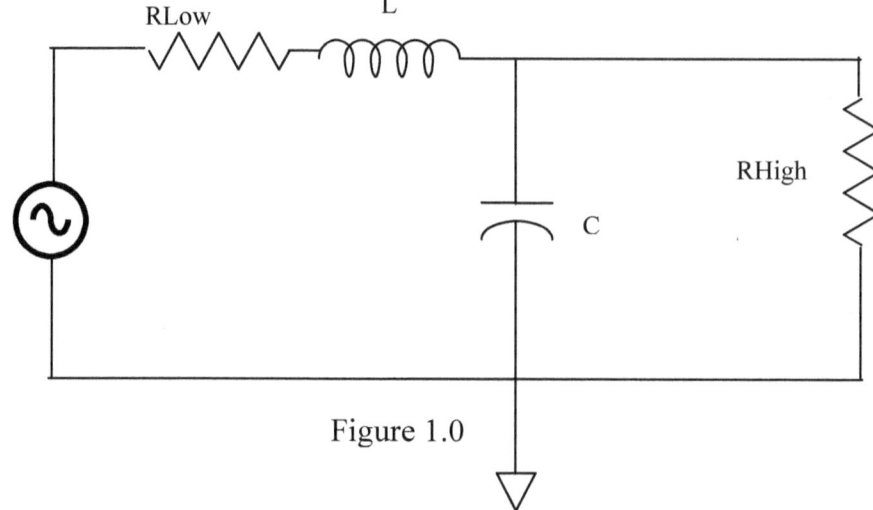

Figure 1.0

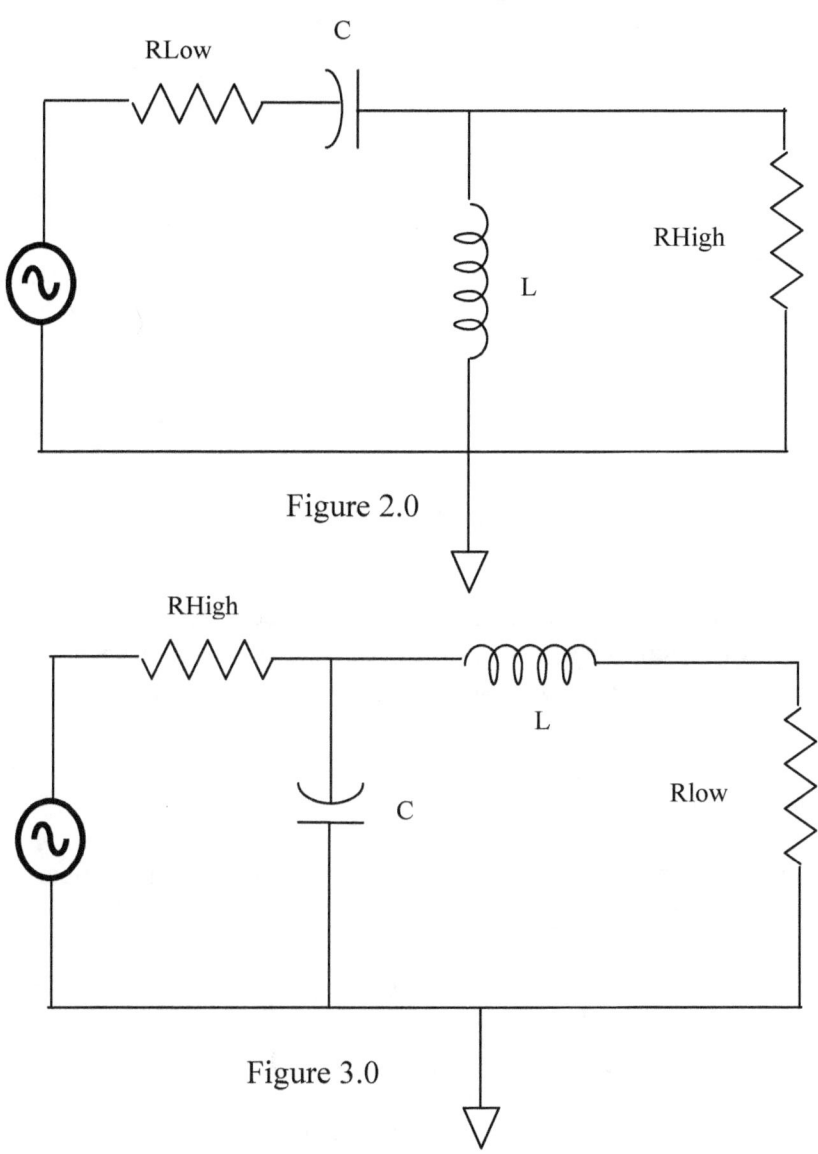

Figure 2.0

Figure 3.0

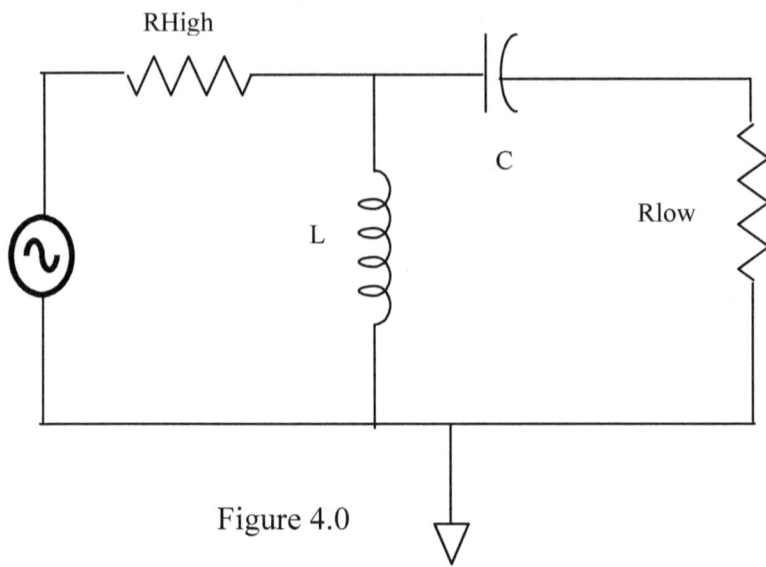

Figure 4.0

By examination of the circuits above, it can be seen that there are *two*
subnetworks in each of the configurations. There is one series subnetwork
and one parallel subnetwork, associated with the respective reactances.

By the law of matching, *these subnetworks must be complex conjugates of*
each other at the frequency of interest.

The Q factors of these subnetworks must be equal at the reference
frequency. In other words:

$$Qs = Qp \qquad (15)$$

where,

$$Qs = Xs/Rlow \qquad (16)$$

$$Qp = Rhigh/Xp \qquad (17)$$

Figure 5 shows an example.

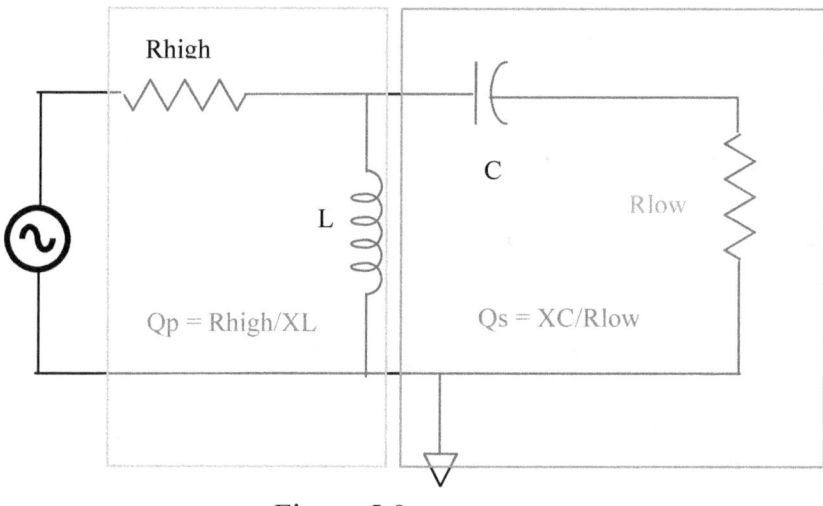

Figure 5.0

In order to find the expressions for use in this technique, we need to find *the equivalency of series and parallel forms.*

Lets assume a series circuit and first write the expression for Q for it:

$$Q = \frac{Xseries}{Rseries} \qquad (18)$$

Then,

$$Xseries = QRseries \qquad (19)$$

The terminal impedance for the series circuit now becomes:

$$\text{Zseries} = \text{Rseries} + j\text{Xseries} = \text{Rseries} + j\text{QRseries} \qquad (20)$$

The conversion of the series impedance into parallel admittance is:

$$\text{Ypar} = \frac{1}{\text{Zseries}} = \frac{1}{\text{Rseries}(1+jQ)} \qquad (21)$$

Separating the real and the imaginary parts:

$$\text{Ypar} = \frac{1}{\text{Rseries}(1+jQ)} \frac{\text{Rseries}(1-jQ)}{\text{Rseries}(1-jQ)} \qquad (22)$$

$$\text{Ypar} = \frac{\text{Rseries}(1-jQ)}{\text{Rseries}^2(1+Q^2)} \qquad (23)$$

$$\text{Ypar} = \frac{\text{Rseries}}{\text{Rseries}^2(1+Q^2)} - \frac{jQ.\text{Rseries}}{\text{Rseries}^2(1+Q^2)} \qquad (24)$$

$$\text{Ypar} = \frac{\text{Rseries}}{\text{Rseries}^2(1+Q^2)} - \frac{j}{\text{Xseries}\ \dfrac{1}{Q^2}(1+Q^2)} \qquad (25)$$

$$\text{Ypar} = \frac{1}{\text{Rpar}} - \frac{j}{\text{Xpar}} \qquad (26)$$

If we now equate the real and imaginary terms of the series and parallel circuit expressions we get:

$$\frac{1}{Rpar} = \frac{1}{Rseries(1+Q^2)} \tag{27}$$

or,

$$\mathbf{Rpar} = Rseries(1 + Q^2) \tag{28}$$

And,

$$\mathbf{Xpar} = \frac{Xseries}{Q^2}.(1+Q^2) \tag{29}$$

Further, the Q is:

$$\mathbf{Q} = \sqrt{\frac{Rpar}{Rseries} - 1} \tag{30}$$

From this result the two sub network Q's can be expressed as:

$$\mathbf{Qs = Qp} = \sqrt{\frac{Rpar}{Rseries} - 1} \tag{31}$$

Of course this can be also expressed in normalized form by dividing the resistor values with the reference resistance (usually 50 Ohms)

Note: Rpar = Rhigh and Rseries = Rlow.

Once the Q is known we can calculate Xs and Xp as:

$$Qs = Xs/Rlow$$

$$Qp = Rhigh/Xp$$

4   Impedance matching of complex terminations.

If the terminations to be matched are complex. Then a modification of the above technique is used. It is best explained using an example. Let the source impedance be $ZS = 20 - j10$ ohms and the load impedance be $ZL = 6 + j12$ ohms. This is illustrated below in Figure L1.0.

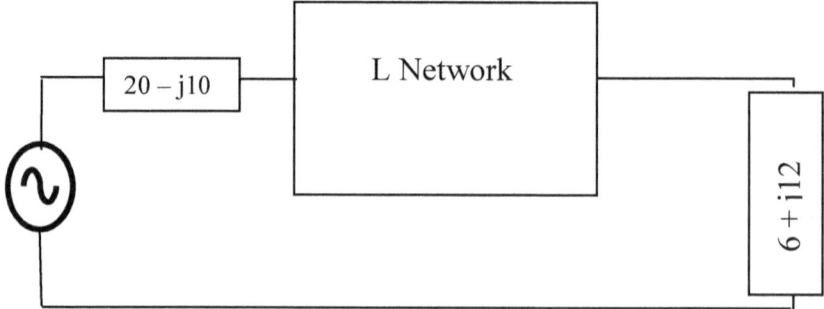

Figure L1.0

Converting the input impedance from a series configuration to a parallel one using the theory above as follows:

$$Q = \frac{10}{20} = 0.5 \qquad\qquad (LC1.0)$$

$$Q^2 = 0.25 \qquad\qquad (LC2.0)$$

Then,

$$\text{Rpar} = 1.25*20 = 25.0 \qquad \text{(LC3.0)}$$

and ,

$$\text{Xpar} = \frac{Xseries}{Q^2}.1+Q^2 \qquad \text{(LC4.0)}$$

or,

$$\text{Xpar} = 50 \qquad \text{(LC5.0)}$$

So the input impedance is transformed to

$$\text{ZS} = 25||\text{-j50}. \qquad \text{(LC6.0)}$$

Here the sign $||$ *stands for "in parallel with".*

A L - section can be used to match the 6 Ohms at the output with the 25 Ohms at the input. Using a type B L – section we get the following circuit. (Figure L2.0)

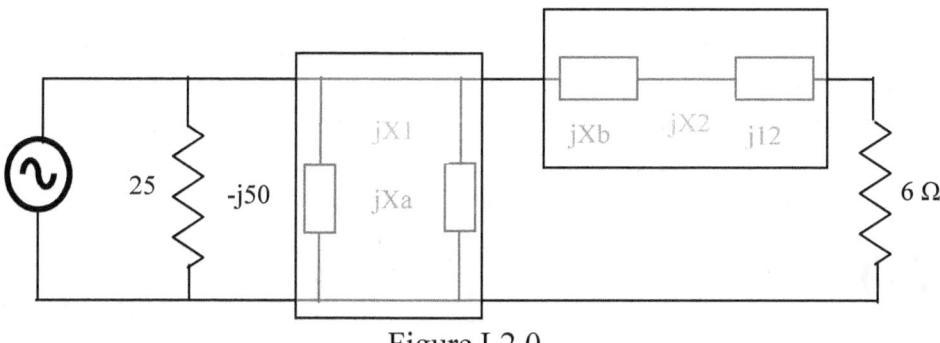

Figure L2.0

The type B matching network reactance, jXa and jXb are shown. If the matching scripts for the type B network are used then the following values are found:

$$X1 = \pm14.05$$
$$X2 = \pm10.68$$

Note that the phase angles define the sign of the reactance and therefore the type of component (L or C). The input reactance can be combined into a new reactance and from that the value of Xa can be found to be:

$$Xa = -19.54 \text{ or } 10.97$$

Similarly the series reactance at the output can be combined and the value of Xb can be found to be:

$$Xb = -1.32 \text{ or } -22.68$$

The dual values are a result of choosing the positive or negative phase angles.

If a type A network is to be used then the output impedance needs to be converted to a parallel form (the reason is that in a type A network the output load is in shunt). This leads to the circuit shown below in Figure L3.0

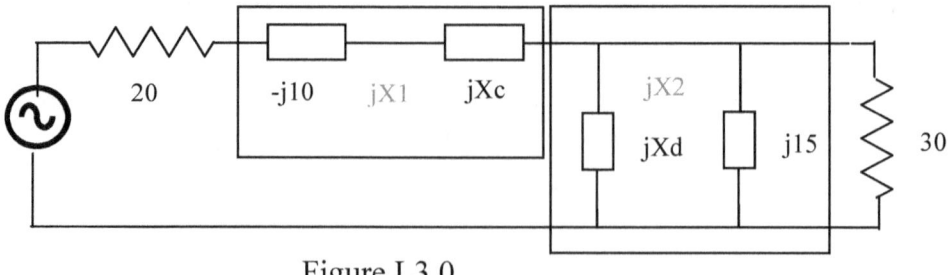

Figure L3.0

Following a similar procedure as in the case of a type B network, we get:

$$X1 = \pm 42.43$$
$$X2 = \pm 14.14$$
$$Xd = -23.20 \text{ or } -11.08$$
$$Xc = -4.14 \text{ or } 24.14$$
$$Xin = 10$$
$$Rin = 20.$$

It should also be noted in passing that there is no reason to assume that all L – Section solutions must exist

In general, impedance matching of complex terminations is achieved by transforming one termination to be the complex conjugate of the other. Other related matching techniques are presented below.

If the load or source has imaginary parts then these can be absorbed in the matching network or eliminated by resonance as shown below.

In order to further the understanding of impedance matching the concept of the *quality factor Q and nodal Q* should be understood.
The underlined Q of a reactive component is defined by:

Qu = Energy stored in the component/Energy dissipated in the component.

When the component is being used in a circuit, a quantity called the loaded Q is defined as follows:

Q_L = Energy stored in the component /Energy dissipated in the component *and the external circuit.*

The <u>nodal Q</u> of a L section matching network is defined as follows. At every node of a L section matching network there is a series impedance Rs + jXs. The nodal Q factor is then defined by:

$Q_N = |Xs|/Rs$.

The nodal Q is also computed as:

$Q_N = \sqrt{[(Rhigh/Rlow)-1]}$ (32)

A. *Absorb the parasitics of the terminations:*

If the Q of the termination is less than the nodal Q (see definition above), then the reactance or susceptance may be absorbed into the matching network. For example consider the circuit below.

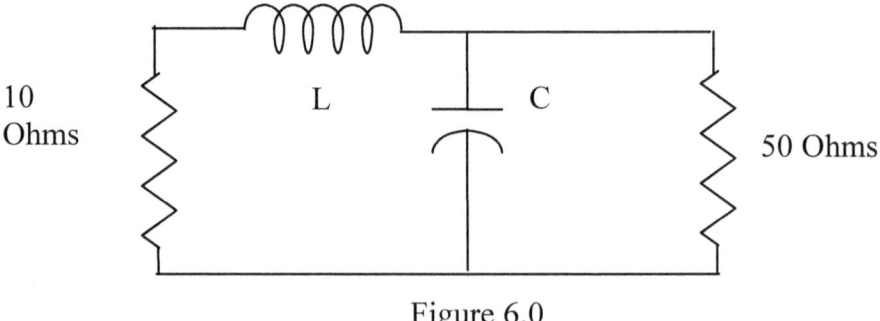

Figure 6.0

In figure 6.0 the nodal Q is:

$$Q_N = 2.0$$

(Please refer to figure 7.0) The total inductor in series with the input cannot exceed 3.18nH. If it is more than that, the Q will exceed 2.0 (the nodal Q). Similarly the capacitance at the output cannot exceed 6.32pF. (This means the total shunt parasitic including Cpara). These limits are set by the ratio of the load resistor and the source resistor by definition of the nodal Q.

As long as the nodal Q value is not exceeded at the output or input the parasitics can be absorbed into the matching circuit. _Note however, that the absorption cannot be on both sides._ When the parasitic inductance _or_ the parasitic capacitance is absorbed, the frequency response is unchanged.

Lparasitic   L = 3.18nH

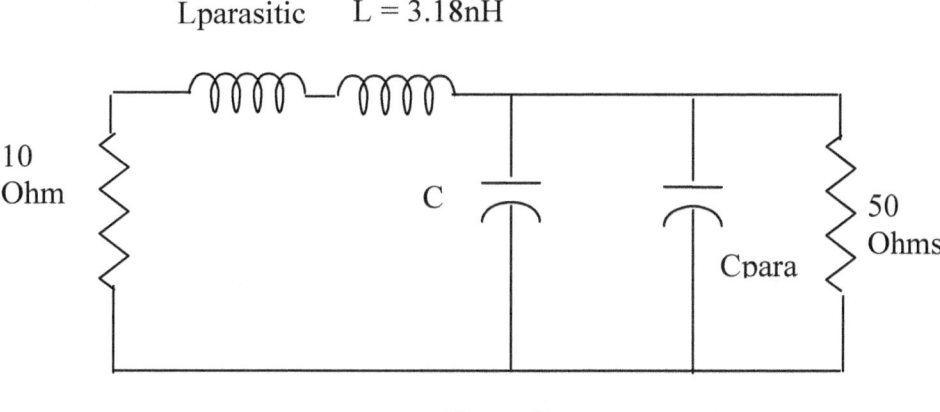

Figure 7.0

B   _Resonate excessive parasitic L or C._

If the parasitic L or C exceeds that maximum allowed value in the absorption technique then a couple of options open up for matching (_at one frequency_).

(B.1) Referring to figure 8.0 below, the capacitance Cmax, that exceeds the maximum allowable capacitance is resonated out by the parallel inductor, LR. This then, leaves the resistive part only visible for matching.

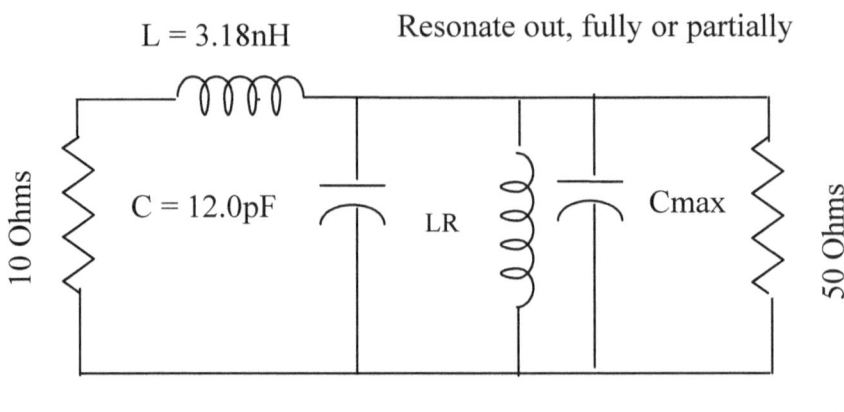

Figure 8.0

The second option is to resonate out a smaller portion of the capacitance Cmax and use the rest to perform the matching,

3.5   Bandwidth considerations:

In all these matching circuits and techniques, bandwidth has not been addressed. Either the matching is at one frequency or over a narrow band of frequencies. In this section the bandwidth is considered as a matching parameter and some techniques are presented for bandwidth control.

Increasing bandwidth: The simplest way to increase the bandwidth of the matching circuitry is to use a two section matching circuit. In other words take the matching L section and break it up into two sections. This is shown below. Choose an impedance Rmid = √(RS * RL)

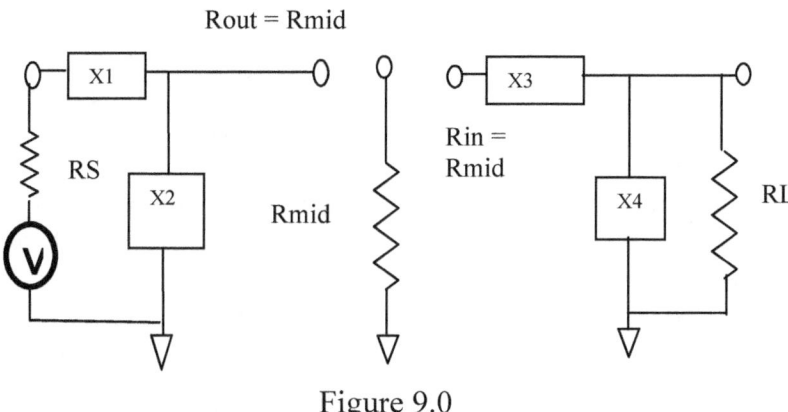

Figure 9.0

Then transform the input impedance to Rmid from RS and transform the output impedance RL to Rmid. This causes the reduction of the termination ratio and the nodal Q for both sections. The net result is an increase in bandwidth.

Optimally Rmid should be between RS and RL geometrically or,

RS/Rmid = Rmid/RL   (33)

A two section matching network actually causes a large change in the bandwidth. However, this does not apply linearly, i.e. a three section network may not bring the same degree of change and subsequent networks bring less and less relative change. *After an increase to about*

6 sections there is no noticeable change. So the practical limit appears to be about 5 sections maximum.

To generalize the intermediate resistance levels we can formulate,

$$Ri = RS^{(k-i)/k}RL^{i/k} \qquad\qquad (34)$$

Here i = number of matching sections and k = interstage index counted from the source side. Again, a practical limit is i = 5.

Note that nodal Q's determine the bandwidth. A high nodal Q will lead to a narrow bandwidth while a low nodal Q will result in a broader bandwidth. From earlier considerations the nodal Q is determined by the _resistance ratio_. Therefore a high resistance ratio will generate a small bandwidth and vice versa. The resistance Rmid is _not a real device_, it is a contrivance to calculate the resistances required only. In reality the input impedance of the second stage represents the load equal to Rmid for the first section and so on. The output impedance of the first section is also Rmid. This acts as the source impedance for the second section. Calculations are based on these considerations.

Decreasing the bandwidth: What if the goal is a decrease in bandwidth? i.e. we do not want a circuit generating spurious responses outside the bandwidth. In this case a similar strategy as above is followed. Again a mid level impedance is used, Rmid. However, in this case Rmid is chosen _outside the range of RS and RL._

The chosen Rmid is dependent on the values of RS and RL in the sense that Rmid can be chosen smaller than or larger than the range of resistances in the circuit.

If the resistances in question are low (say 100 Ohm match to 50 Ohm) then obviously choosing a higher resistance (Rmid) can cause problems so the choice should be to go for a lower (< 50 ohm) Rmid. If the range of matching is 15 ohms to 50 ohms then a choice of Rmid greater than 50 Ohms is a better choice since matching with low impedance levels (a few ohms) might be difficult. To show these strategies, consider the two circuits presented below.

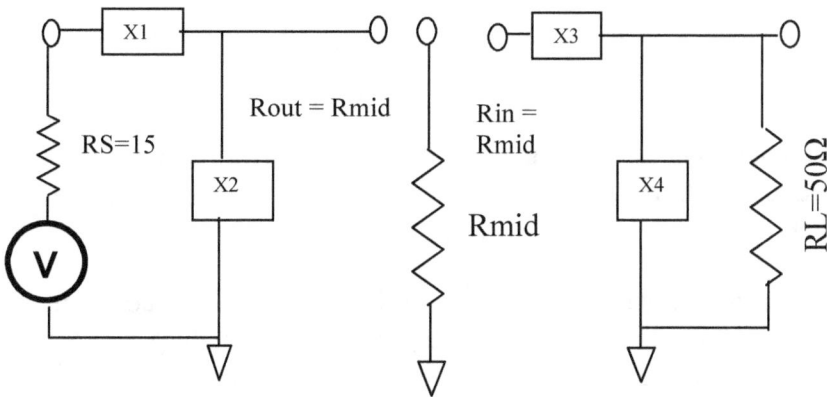

Figure 10.0 In this circuit Rmid is chosen lower than 15 Ohms.

In Figure 10.0, Rmid is chosen below 15 Ohms. Why? In order to answer this question it is best to use an example. Figure 11.0 shows a matching circuit using a low pass first stage and a high pass second stage.

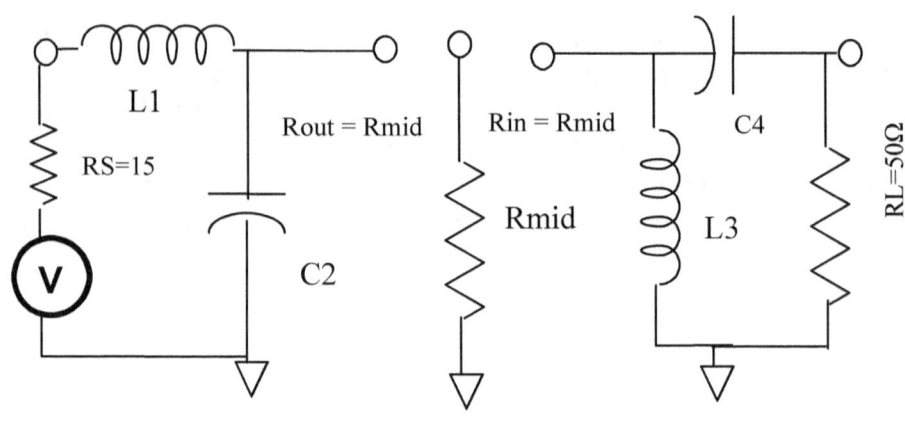

Figure 11.0

There appears to be no analytical way of determining the Rmid. It has to be chosen. In the case of Figure 11.0 a value of 250 Ohms is chosen. Then:

The nodal Q for the first LC circuit is 15.7. For the second circuit it is 4. Obviously the bandwidth will be governed by the larger of the two. Then L1 and C2 can be determined by the methods described before.

A dual of this circuit is the high pass - low pass combination. The treatment is the same.

Part IV

Transmission line matching circuits:

4.1 Simple cascaded line matching:

At elevated frequencies, transmission line matching becomes practical. The simplest match with a transmission line is given by adding a length of line with characteristic impedance Zo, between the resistive load and resistive source.

Characteristic impedance of the line Zo = √(RL.RS)   (35)
And the electrical length is:

θ = 90 Degrees (Note electrical lengths of 270 Degrees, 450 Degrees etc are also a solution)

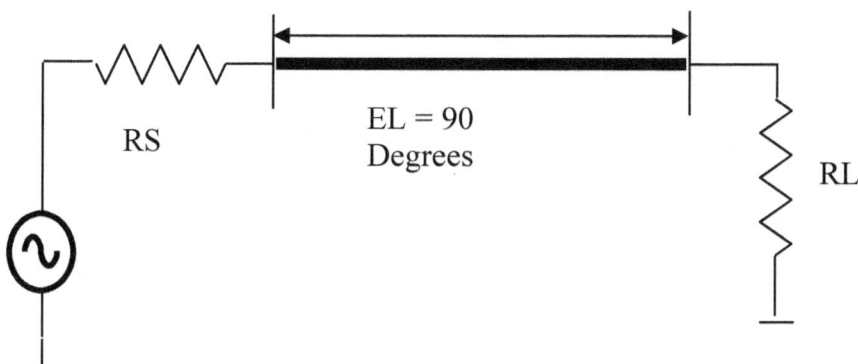

4.1.1   <u>Definition of Electrical length:</u>

As a note, the electrical length of a transmission line is defined by the following:

1.0 The wave number or phase constant = $\beta = 2\pi/\lambda$

2.0 The electrical length is defined by $\theta = \beta l$ where l = physical length

3.0 $\theta = \beta l = (l/\lambda) *360$ degrees

Here λ is the wavelength of the signal in the applicable dielectric (sometimes called the guide wavelength).

4.0 For frequencies in Ghz, [360 * fGhz * l(cm) * $\sqrt{\varepsilon eff}$]/30 cm

In this case frequency is in Ghz, physical length is in centimeters.

For example:

Let frequency be 1 Ghz.
Let $\lambda = 0.8 \lambda$(air) or $\sqrt{\varepsilon eff} = 1.25$
Let $l = 0.1$ meters $= 0.1E2$ centimeters

Then :

$\theta = [360* l*0.1E2*1.25]/30$ degrees

$\theta = 150$ degrees

4.1.2   Definition of β: Sometimes β is referred to as the phase constant of the line or guide. If the cartesian coordinate system is used and a coordinate, say "z" is used as the direction of wave propagation then βz measures the instantaneous phase at point z on the line with respect to z =0.

In addition, voltage or current on the line is the same at any two points separated in z such that βz differs by multiples of 2π. Since the shortest distance between points where voltage or current is at the same phase is a *wavelength*, then:

$$\beta\lambda = 2\pi$$

(replacing z by λ),

$$\beta = 2\pi/\lambda$$

Complex impedances can also be matched with a section of transmission line. The only stipulation is that RS should not be equal to RL. If this is true then the characteristic impedance of the line is given by:

$$\sqrt{\{[(RS^2 + XS^2)RL - (RL^2 + XL^2)RS]/(RS - RL)\}} \qquad (36)$$

where, the source impedance is RS+jXS and the load impedance is

RL+jXL.

and the electrical length is:

$$\theta = \tan^{-1}[Zo(RL-RS)/(XSRL - XLRS)] \qquad (37).$$

4.2 The quarterwave transformer:

The quarter wave transformer is a quarter wavelength of transmission line with a characteristic impedance of Zi placed between a transmission line of characteristic impedance Zo and a real load impedance of RL. Figure 12.0 below depicts the circuit.

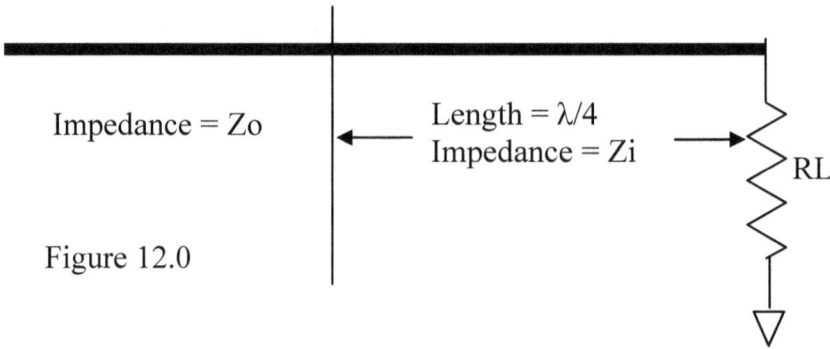

Impedance = Zo

Length = $\lambda/4$
Impedance = Zi

RL

Figure 12.0

Obviously the impedance is dependent on the width of line and the thickness of the insulation on which it resides (if it is a microstrip line). In any case, transmission lines can be any style. The characteristic impedance can usually be calculated fairly easily.

The analytic expressions associated with this type of matching network are presented below.

The input impedance of a transmission line terminated in a load ZL is given by:

$$\text{Zin} = Z1\frac{ZL + jZ1\tanh(\gamma l)}{Z1 + jZL\tanh(\gamma l)} \tag{38}$$

Here,

>   Z1 = characteristic impedance of the line
>   ZL = load
> γ = propagation constant

If the load is real and equal to RL,

Then,

$$\text{Zin} = Z1\frac{RL + jZ1\tanh(\gamma l)}{Z1 + jRL\tanh(\gamma l)} \tag{39}$$

If the length of the transmission line is a *quarter wavelength*, then the input impedance is purely real.

Then,

$$\gamma l = \frac{2\pi}{\lambda}\cdot\frac{\lambda}{4} = \frac{\pi}{2} \tag{40}$$

> λ = wavelength of the signal

The terms tanh() become unbounded. Taking the limit we get:

$$\text{Zin} = \frac{Z1^2}{RL} \tag{41}$$

If the system is matched then the input impedance must be equal to the characteristic impedance Zo. Or,

$$Zin = \frac{Z1^2}{RL} = Zo \qquad (42)$$

Which leads to:

$$Z1 = \sqrt{(ZoRL)} \qquad (43)$$

For example: Match a 50 ohm line to a 100 ohm load. Determine the characteristic impedance of the matching quarterwave line.

Here RL = 100.
Zo = 50. Then:
$Z1 = \sqrt{(100*50)} = 70.71$ Ohms.

This is one of the simplest ways to match a load to the line.

4.3   Frequency response of the quarter wave transformer.

The frequency response of the reflection coefficient is given by:

$$\Gamma(\omega) = \frac{Zin(\omega) - Zo}{Zin(\omega) + Zo} \qquad \text{f1.0}$$

$$Zin(\omega) = Z1 \frac{ZL + jZ1 \tanh(\beta(\omega)l}{Z1 + jZL \tanh(\beta(\omega)l} \qquad \text{f2.0}$$

$$Z1 = \sqrt{(ZoZL)} \qquad\qquad\qquad\qquad\qquad\qquad\qquad f3.0$$

$$\beta(\omega)l = \frac{2\pi}{\lambda} \cdot \frac{\lambda o}{4} = \frac{\pi}{2}\left(\frac{f}{fo}\right) \qquad\qquad\qquad\qquad f4.0$$

f = frequency variable,
fo = reference frequency, design center frequency for the quarter wave transformer.

4.4 Multisection matching transformer:

4.41 The binomial transformer
A multisection matching transformer is made by connecting K transmission lines in series between the feeder line of characteristic impedance Zo and the load ZL.

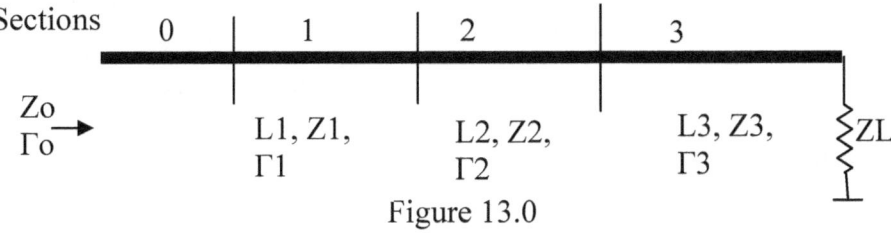

Figure 13.0

Where , Li = line lengths, Zi = impedances, Гi = reflection coefficients for each section of line.

Note that the reflection coefficient is defined by,

$$\Gamma = \frac{Z - Zo}{Z + Zo} \qquad\qquad\qquad\qquad (44)$$

where Zo is the source impedance and Z is the load impedance.

Thus for section 0 above we have a reflection coefficient of:

$$\Gamma o = \frac{Z1 - Zo}{Z1 + Zo} \tag{45}$$

for any n except N we have:

$$\Gamma n = \frac{Z_{n+1} - Z_n}{Z_{n+1} + Z_n} \qquad [\,1 \leq n \leq N\text{-}1] \tag{47}$$

and for the last section:

$$\Gamma N = \frac{ZL - ZN}{ZL + ZN} \tag{48}$$

We now make each section of the line the _same length_. In this case the reflection coefficients are of the same sign. Under these assumptions the total reflection coefficient may be written as:

$$\Gamma(\theta) = \Gamma o + \Gamma 1 e^{-2j\theta} + \Gamma 2 e^{-j4\theta} + \ldots\ldots + \Gamma N e^{-j2N\theta} \tag{49}$$

An additional assumption made at this stage is that the reflection coefficients are symmetric. This implies that $\Gamma o = \Gamma N$, $\Gamma 1 = \Gamma N\text{-}1$ and so on.

Equation (49) leads to the following summation:

$$\Gamma(\theta) = A \sum_{n=0}^{n=N} Cne^{-j2n\theta} \tag{50}$$

$$\theta = \frac{\pi}{2} \frac{f}{fo} \tag{50.1}$$

Cn is the binomial coefficient and A is the amplitude factor. The binomial coefficient is given by:

$$Cn = \frac{N!}{n!(N-n)!} \tag{51}$$

Equating the coefficients Cn and the reflection coefficients leads to:

$$\Gamma n = ACn \tag{52}$$

It can be shown that the amplitude factor is given by:

$$A = 2^{-N} \frac{ZL - Zo}{ZL + Zo} \tag{53}$$

Furthermore, the adjacent sections of the binomial transformer are related by:

$$\Gamma n = ACn = \frac{Z_{n+1} - Z_n}{Z_{n+1} + Z_n} \tag{54}$$

or,

$$Z_{n+1} = Z_n \frac{1 + ACn}{1 - ACn} \tag{55}$$

It can also be shown that:

$$Z_{n+1} = Zo * \exp(K) \tag{56}$$

_____

where K is:

$$\sum C_k 2^{-N} \ln(\frac{ZL}{Zo}) \tag{57}$$

The summation is from k = 0 to n.

To further clarify these concepts an example is presented below.

Let us assume we want to design a 3 – section binomial matching transformer to match a 100 ohm load to a 50 ohm line and find the percentage bandwidth for an overall reflection coefficient of 0.3.
Solution:

Step 1.0   Find the binomial coefficients from the identity Cn = $\frac{N!}{n!(N-n)!}$.

Thus:

$$C0 = \frac{3.2.1}{1.3.2.1} = 1$$

$$C1 = \frac{3.2.1}{1.2.1} = 3$$

$$C2 = \frac{3.2.1}{2.1.1} = 3$$

$$C3 = \frac{3.2.1}{3.2.1.1} = 1$$

Note: 0! (factorial 0) is actually equal to 1.0

Step 2.0   Calculate ln[ZL/Zo] = ln(100/50]= ln2 = 0.6931.

Step 3.0   Calculate equation (57)

$$\frac{\sum C_k * 0.6931}{8.0} = \sum C_k * 0.0866 \qquad (58)$$

or,

$$Zn+1 = 50*0.0866* \sum C_k \quad (\text{ summation is from } k=0 \text{ to } n)$$

Therefore,

$Z1 = 4.330*1 = 4.330$ Ohms , here $\sum C_k = C0$ since $n = 0$
$Z2 = 4.330 * 4 = 17.32$ Ohms , here $\sum C_k = C0 + C1$ since $n = 1$
$Z3 = 4.330 * 7 = 30.31$ Ohms , here $\sum C_k = C0 + C1 + C2$ since $n = 2$

Note that all lengths are quarter wave long.

4.4.2   The percentage bandwidth for the binomial transformer:

The percentage bandwidth is given by the following expression:

$$\frac{\Delta f}{fo} = 2 - (\frac{4}{\pi}) \cos^{-1}[0.5 * (\frac{\Gamma_m}{|A|})^{\frac{1}{N}}] \qquad (59)$$

and ,

$$A = 2^{-N} \frac{ZL - Zo}{ZL + Zo} \qquad (59.1)$$

Γ_m = overall reflection coefficient at its maximum allowable value, *a specified value between 0 and 1 obviously.*

Again using the expression (59) and (59.1) above, the percentage bandwidth can be calculated. The example shows the value for the parameters chosen there.

The percentage bandwidth for the example above is:

$$A = \frac{1}{8}(\frac{50}{150}) = 0.04 = \frac{1}{24}$$

$$Bw\% = 2.0 - (1.27)\cos^{-1}[0.5*(\frac{2.4}{1})^{0.33}] = 0.72 \ (72\%)$$

4.4.3 Multisection matching with a Chebyshev transformer:

A still broader bandwidth can be achieved with a multisection circuit using impedances based on Chebyshev functions. The trade-off is the ripple in the passband. However, even in this case we can still specify a certain maximum allowable reflection coefficient. The following describes the design.

Chebyshev polynomials have the following properties:

1.0   Even ordered Chebyshev polynomials are even functions.
2.0   Odd ordered Chebyshev polynomials are odd functions.
3.0   The magnitude of any Chebyshev polynomial is unity or less than unity in the range of $-1 \le x \le 1$ where x is the independent variable of the polynomial.
4.0 $Tn(1) = 1$ for all Chebyshev polynomials.
5.0   All the roots of Chebyshev polynomials lie in the range of $-1 \le x \le 1$.

Chebyshev polynomials up to the ninth order are listed below

$T0(x) = 1$
$T1(x) = x$
$T2(x) = 2x^2 - 1$
$T3(x) = 4x^3 - 3x$
$T4(x) = 8x^4 - 8x^2 + 1$
$T5(x) = 16x^5 - 20x^3 + 5x$
$T6(x) = 32x^6 - 48x^4 + 18x^2 - 1$
$T7(x) = 64x^7 - 112x^5 + 56x^3 - 7x$
$T8(x) = 128x^8 - 256x^6 + 160x^4 - 32x^2 + 1$
$T9(x) = 256x^9 - 576x^7 + 432x^5 - 120x^3 + 9x$

In order to facilitate design it is useful to write the Chebyshev polynomials in terms of $x = \cos\theta$. Once this is done, the transformed polynomials may be written:

$T0(\theta) = \cos(0) = 1.0$
$T1(\theta) = \cos(\theta) = \cos(\theta)$
$T2(\theta) = \cos(2\theta) = 2\cos^2(\theta) - 1$
$T3(\theta) = \cos(3\theta) = 4\cos^3(\theta) - 3\cos(\theta)$
$T4(\theta) = \cos(4\theta) = 8\cos^4(\theta) - 8\cos^2(\theta) + 1$
$T5(\theta) = \cos(5\theta) = 16\cos^5(\theta) - 20\cos^3(\theta) + 5\cos(\theta)$
$T6(\theta) = \cos(6\theta) = 32\cos^6(\theta) - 48\cos^4(\theta) + 18\cos^2(\theta) - 1$
$T7(\theta) = \cos(7\theta) = 64\cos^7(\theta) - 112\cos^5(\theta) + 56\cos^3(\theta) - 7\cos(\theta)$
$T8(\theta) = \cos(8\theta) = 128\cos^8(\theta) - 256\cos^6(\theta) + 160\cos^4(\theta) - 32\cos^2(\theta) + 1$
$T9(\theta) = \cos(9\theta) = 256\cos^9(\theta) - 576\cos^7(\theta) + 432\cos^5(\theta) - 120\cos^3(\theta) + 9\cos(\theta)$

or a compact notation is:

$$\Gamma n(\cos\theta) = \cos n\theta \tag{60}$$

Also Chebyshev polynomials for all arguments may be written as,

Tn(x) = cos(ncos-1x)   for |x| < 1   (61)
Tn(x) = cosh(ncosh-1x)   for |x| > 1   (62)

In order to use Chebyshev polynomials, the end points of the passband given by $(\theta_m, \pi - \theta_m)$ with the center frequency at $\pi/2$, must be mapped onto the range where the Chebyshev polynomials satisfy the identity,

$\quad$ | Tn(cosθ)|≤ 1.

Such a mapping exists and is given by:

$$ Tn\left(\frac{\cos\theta}{\cos\theta_m}\right) = Tn(\sec\theta_m\cos\theta) $$

Using this mapping the polynomials can be written as:

$\quad$ $T0(\sec\theta_m\cos\theta) = 1$

$\quad$ $T1(\sec\theta_m\cos\theta) = \sec\theta_m\cos\theta$,

and so on.

The reflection coefficient of a N multisection transformer is given by:

$$ \Gamma(\theta) = 2\,e^{-jN\theta}\left[\Gamma_0\cos N\theta + \Gamma_1\cos(N-2)\theta + ... + \Gamma_n\cos(N-2n)\theta + ...\right] \text{ (62.1)} $$

which reduces to,

$$ \Gamma(\theta) = Ae^{-jN\theta}\,\Gamma_N(\sec\theta_m\cos\theta) \tag{63} $$

The maximum magnitude of the reflection coefficient in the passband is A.

This is because the maximum magnitude of the Chebyshev polynomial cannot exceed unity in the passband..
The value of A is given by:

$$A = \frac{\ln(\frac{ZL}{Zo})}{2\Gamma_N(\sec\theta_m)} \qquad (64)$$

(This result is derived by taking the limit of the reflection coefficient T(θ) as θ approaches zero. Also because $\Gamma(0) = \frac{ZL - Zo}{ZL + Zo}$)

Using this equation we can also write for the reflection coefficient:

$$\Gamma(\theta) = \frac{\ln(\frac{ZL}{Zo})}{2\Gamma_N(\sec\theta_m)} \cdot e^{-jN\theta} \, \Gamma_N(\sec\theta_m \cos\theta) \qquad (65)$$

from equations (63) and (64).

The angle θ_m can be found from the expression below (presented without proof).

$$\sec(\theta_m) = \cosh\left[\frac{1}{N}\cosh^{-1}\left|\frac{\ln(ZL/Zo)}{2\Gamma_m}\right|\right] \qquad (66)$$

The characteristic impedances can be found from,

$$Z_{n+1} = Z_n e^{2\Gamma_m} \qquad (67)$$

___

where the Γ_m are the reflection coefficients.
The fractional bandwidth is,

$$\frac{\Delta f}{f_o} = 2 - \frac{4\theta_m}{\pi} \tag{68}$$

The following example is presented to clarify these design concepts.
A 4 section Chebyshev transformer is required to match a 300 Ohm load to
a 50 Ohm line. The maximum reflection coefficient cannot exceed 0.1.

From eqn (66) we get,

$$\sec \theta_m = \cosh\left[\frac{1}{4}\cosh^{-1}\left|\frac{\ln(6)}{2(0.1)}\right|\right] \quad \text{which gives } \theta_m = 38.2 \text{ Degrees.} \tag{69}$$

The reflection coefficient for a fourth order section is given by:

$$\Gamma(\theta) = Ae^{-j4\theta} T_4(\sec\theta_m \cos\theta) \tag{70}$$

Now we can replace T_4 by its expanded version:

$$T_4(\sec\theta_m \cos\theta) = \sec^4 \theta_m (\cos 4\theta + 4\cos 2\theta + 3) - 4\sec^2 \theta_m (\cos 2\theta + 1) + 1 \tag{71}$$

Also the equation for the reflection coefficient with $N = 4$ is:

$$\Gamma(\theta) = 2 e^{-j4\theta} \left[\Gamma_0 \cos 4\theta + \Gamma_1 \cos 2\theta + .\Gamma_2 / 2\right] \tag{72}$$

Now if equation (71) is substituted in equation (70) and after some
manipulation the coefficients of the cosine terms are equated, we get:

$$\Gamma_o = 0.131$$
$$\Gamma_1 = 0.201$$
$$\Gamma_2 = 0.239$$

Now additionally, because of symmetry:

$$\Gamma_4 = \Gamma_0$$
$$\Gamma_3 = \Gamma_1$$

Therefore, using equation (67) we get;

$$Z_1 = Z_0 e^{2\Gamma_0} = 65 \text{ Ohm}$$

Similarly,

$Z_2 = 97.1$	Ohm
$Z_3 = 156.7$	Ohm
$Z_4 = 234.2$	Ohm

Transmission line expressions and formulas

4.5.1 Transmission line facts:

Before going any further in this discussion the following facts about transmission lines should be considered:

0.0 Transmission lines that have electrical lengths *less than 90 Degrees* behave *inductively for short circuited loads* and *capacitively for open circuited loads.*

1.0 A *RF short circuit* can be produced *at any point* in a circuit by using a *short circuited transmission line with a half wavelength electrical length*. This short repeats itself every multiple of a half wavelengh.

2.0 A transmission line's input impedance is always equal to the termination at adjacent ends of the line if the characteristic impedance of the line is *the same as* the termination.

3.0 Transmission lines have *large transformation capabilities*. A short circuit can be transformed to an open circuit by using a *90 Degree* long line. The same is true of an open circuit. An open circuit can be transformed into a short circuit by a *90 Degree* line.

4.0 Cascaded transmission lines form *concentric circles* on a normalized Smith Chart if the load impedance is normalized to the characteristic impedance of the transmission line.

5.0 Parallel open and short circuited stubs behave *inductively or capacitively* as long as their electrical length is *less than 90 Degrees*.

6.0 Parallel stubs *always* move on the constant conductance circles on the Smith Chart.

7.0 A correctly selected combination of a *parallel stub and cascade transmission line* can be used to *transform any point* on the Smith Chart to any other point. The topology of this combination is dependent on the relationship of the two points. Sometimes the cascade line can be used first followed by the stub, while at other times the stub is followed by the cascaded line.

4.5.2 Electrical length

Sooner or later, the design engineer who is working in microwave or high frequency electronics, is going to come up against the concept of electrical length. In order to understand this concept lets work out the following arithmetic:

1.0 The wave number* or phase constant $= \beta = 2\pi/\lambda$

2.0 The electrical length is defined by $\theta = \beta l$ where l = physical length.

3.0 $\theta = \beta l = (l/\lambda) *360$ degrees

Here λ is the wavelength of the signal in the applicable dielectric (or sometimes called the guide wavelength).

4.0 For frequencies in Ghz, this becomes: $[360 * fGhz * l(cm) * \sqrt{\varepsilon eff}]/30$ cm

In this case frequency is in Ghz, physical length is in centimeters.

For example:

Let frequency be 1 Ghz.

Let $\lambda = 0.8\ \lambda(air)$ or $\sqrt{\varepsilon eff} = 1.25$

Let $l = 0.1$ meters $= 0.1E2$ centimeters.

Then :

$\theta = [360* 1*0.1E2*1.25]/30$ degrees

- See the Signal Processing Group Inc. blog located at: **http://signalpro.biz/wordpress**

Transmission line parameters and characterization

In this section we present some parameters of transmission lines useful to the practicing engineer or student of transmission line matching.

The Propagation Constant:

A traveling wave on a transmission line has a voltage v and current i. These two quantities are related by the *Characteristic Impedance* of the line as:

$$Zo = \frac{v}{i} = \sqrt{\frac{Ro + j\omega Lo}{Go + j\omega Co}}$$ (S24.0)

where:

Ro = resistance per unit length of the line.

Go = shunt conductance per unit length of the line.

Lo = Series inductance per unit length of the line.

Co = shunt capacitance per unit length of the line.

The equivalent RLCG circuit of a transmission line is shown below.

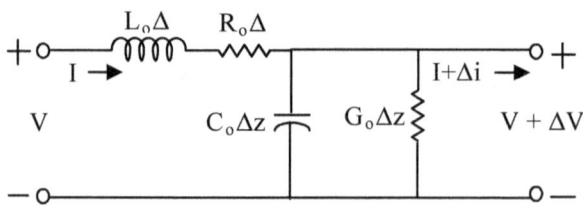

The propagation constant of a real, lossy transmission line is given by:

$$\gamma = \alpha + j\beta \qquad\qquad (S25.0)$$

which is also given by:

$$\sqrt{[(Ro+j\omega Lo)(Go+j\omega Co]} \qquad\qquad (S26.0)$$

Here,

α = attenuation constant in nepers/unit length

β = phase constant in radians/unit length

For the ideal line the propagation constant may also be written:

$$\gamma = j\omega\sqrt{(LoCo)} \qquad\qquad (S26.1)$$

For <u>low loss lines,</u>

$$\alpha = \frac{Ro}{2Zo} + \frac{GoZo}{2} \qquad\qquad (S26.2)$$

$$\beta = \omega\sqrt{LoCo}\left\{1.0 - \left[\frac{RoGo}{4\omega^2 LoCo}\right] + \left[\frac{Go^2}{8\omega^2 Co^2}\right] + \left[\frac{Ro^2}{8\omega^2 Lo^2}\right]\right\} \qquad (S26.3)$$

For <u>the ideal line,</u>

$\alpha = 0.$

$$\beta = \omega\sqrt{(LoCo)} = \frac{\omega}{v} = \frac{2\pi}{\lambda} \qquad \text{(S26.4)}$$

Phase velocity:

The phase velocity is given by:

$$v = \frac{\omega}{\beta} \qquad \text{(S27.0)}$$

Characteristic impedance:

If the lossy components of the line are essentially zero, i.e. the line is considered lossless, then the line characteristic impedance can be defined as:

$$Zo = \sqrt{\frac{Lo}{Co}} \qquad \text{(S28.0)}$$

or generally,

$$Zo = \sqrt{\frac{Ro + j\omega Lo}{Go + j\omega Co}} \qquad \text{(S28.1)}$$

For low – loss lines,

$$Zo = \sqrt{\frac{Lo}{Co}} \left\{ 1.0 + j \left[\left(\frac{Go}{2\omega Co} \right) - \left(\frac{Ro}{2\omega Lo} \right) \right] \right\} \qquad \text{(S28.2)}$$

Electrical length:

This term refers to the ratio of the physical length 'l' of the transmission line to the wavelength, λ in the applicable dielectric. The wavelength in the applicable dielectric, sometimes called the guide wavelength is given by:

$$\lambda_G = \frac{\lambda_o}{\sqrt{\varepsilon_{eff}}} \qquad \text{(S29.0)}$$

Here:

ε_{eff} = Effective dielectric constant , and λo = wavelength in air

If the signal wave propagates in homogeneous media then the effective dielectric constant is equal to ε_R, the relative dielectric constant. However, in cases of microstrip line, for instance, where part of the wave is in the air and part in the dielectric, the effective dielectric constant is *not equal* to the relative dielectric constant. In such a case the effective dielectric constant has to be calculated.

Fractional wavelength:

Fractional wavelength is the ratio of the physical length of the line to the effective or guide wavelength.

$$\text{Fractional wavelength (\%)} = \left[\frac{l}{\lambda_G}(100) \right] \qquad \text{(S30.0)}$$

or,

In degrees:

$$\theta(\text{Degrees}) = [l/\lambda_G] [360] \qquad \text{(S31.0)}$$

Please note that in the above equations, the quantity l can be confused with the numeral 1. Thus l/λ_G is length/guide wavelength, not one divided by guide wavelength.

Input impedance:

An ideal transmission line terminated with a load of ZL has an input impedance of:

$$ZIN = Zo \frac{ZL + jZo\tan(\theta)}{Zo + jZL\tan(\theta)} \qquad \text{(S32.0)}$$

For a general transmission lines the input impedance expression is:

___

$$ZIN = Zo \frac{ZL + jZo \tanh(\gamma l)}{Zo + jZL \tanh(\gamma l)} \qquad \text{(S33.0)}$$

where ZL = the load or termination and Zo is the characteristic impedance.

Impedance of a shorted line:

This is also addressed in the section on stubs. In any case, the impedance of a *ideal shorted line* is: ($\theta = \beta l$, the electrical length)

$$Z = jZo\tan(\theta) \qquad \text{(S34.0)}$$

while for *a general line:*

$$Z = Zo\tanh(\gamma l) \qquad \text{(S35.0)}$$

and for *a low-loss line:*

$$Z = Zo \frac{\alpha l + j \tan \theta}{1.0 + j\alpha l \tan \theta} \qquad \text{(S36.0)}$$

Impedance of an open line:

For an *ideal line:*

$$Z = -jZoCot\theta \qquad \text{(S37.0)}$$

For a *general line,*

$$Z = ZoCoth(\gamma l) \tag{S38.0}$$

For a *low – loss line:*

$$Z = Zo\frac{1 + j\alpha l \tan\theta}{\alpha l + j\tan\theta} \tag{S39.0}$$

Impedance of a quarter-wave line:

For an *ideal line:*

$$Z = \frac{Zo^2}{ZL} \tag{S40.0}$$

For a *general line,*

$$Z = Zo\frac{ZL + Zo\coth\alpha l}{Zo + ZL\coth\alpha l} \tag{S41.0}$$

For a *low – loss line:*

$$Z = Zo\frac{Zo + ZL\alpha l}{ZL + Zo\alpha l} \tag{S42.0}$$

Impedance of a half-wave line:

For *an ideal line:*

$$Z = ZL \tag{S43.0}$$

For *a general line:*

$$Z = Zo\frac{ZL + Zo\tanh\alpha l}{Zo + ZL\tanh\alpha l} \tag{S44.0}$$

For a *low-loss line:*

$$Z = Zo\frac{ZL + Zo\alpha l}{Zo + ZL\alpha l} \tag{S45.0}$$

Voltage along the line:

For the *ideal line:*

$$Vline = Vinput.Cos(\beta z) - jIinput.ZoSin(\beta z) \tag{S46.0}$$

where:

Vinput is the input voltage and,

Iinput is the input current.

For a *general line*:

$$Vline = VinputCosh(\gamma z) - Iinput.Zo.Sinh(\gamma z) \tag{S47.0}$$

Current along the line:

For the ideal line:

$$\text{Iline} = Iinput.Cos(\beta z) - j\frac{Vinput}{Zo}Sin(\beta z) \qquad\qquad (S48.0)$$

For the general line:

$$\text{Iline} = Iinput.Cosh(\gamma z) - \frac{Vinput}{Zo}.Sinh(\gamma z) \qquad\qquad (S49.0)$$

The reflection coefficient:

For both an *ideal line* and a *general line*:

$$\Gamma = \frac{ZL - Zo}{ZL + Zo} \qquad\qquad (S50.0)$$

The standing wave ratio:

For both the ideal line and a general line,

SWR = [1+| Γ|]/[[1-| Γ|]   (S51.0)

Microstrip lines.

Microstrip lines are used commonly both in IC design, and in PCB design. Some interesting features of these types of transmission lines are described below. <u>For a really helpful script use *txline.exe.* Also read *Ref. 5.*</u>

A cross section of the microstrip line is shown below. As shown, it consists of a top conducting metallic strip, an insulating substrate and a ground plane on the bottom of the substrate.

Once the Zo of the line is known most of the above formulas can be used in synthesis and analysis of the impedance matching networks.

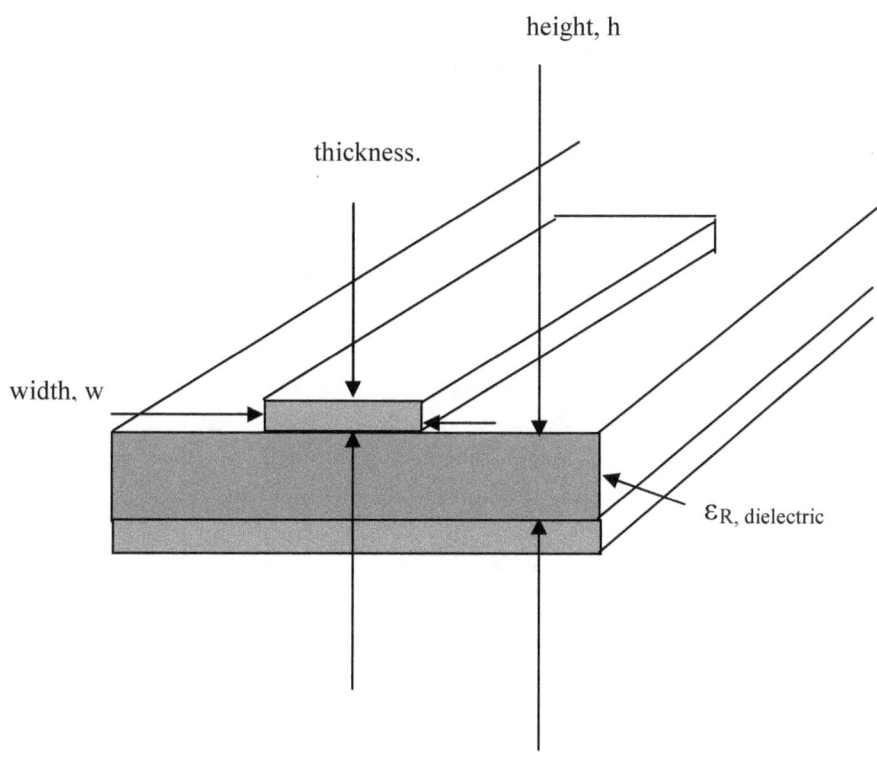

height, h

thickness.

width, w

ε_R, dielectric

Series stubs or short high impedance (read narrow) lengths of line are inductive and wide short lengths of microstrip are capacitors.

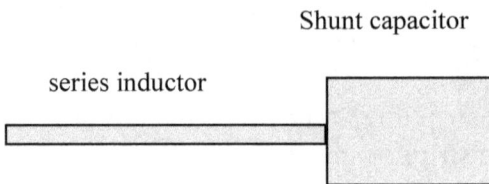

Shunt capacitor

series inductor

In addition a gap in the microstrip line can be assumed to be a series capacitor. This is shown below.

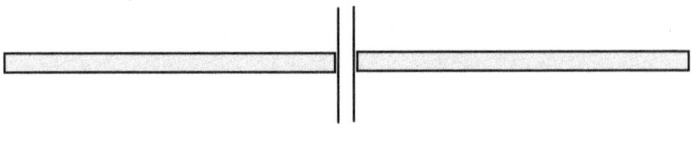

Series gap, series capacitor

Other type of components using microstrip lines have been described above.

Part V
The Smith Chart and impedance matching techniques

The Smith Chart is a valuable aid in graphically visualizing the reflection coefficient and Y-Z networks and many other associated quantities. It was invented by Phillip Smith in 1932 and it is still being used today. Its principal operations and techniques are discussed in this part of the eCADbook. Impedance matching techniques using it are also presented.* *The reader should use the freeware mentioned in the text and use it to clearly see the relationships on the chart in color. This book is printed in black and white and some of the graphic/descriptive information may not be too clear on the examples provided.**

The Smith Chart is a mapping between the normalized *complex impedance plane* given by z = r+jx and *the complex reflection plane*. The normalized impedance is given by,

$$z = Z/Zo = [R + jX]/Zo \qquad (S0.0)$$

The right hand side of a *normalized* complex impedance plane is such that the values of an impedance that has a real impedance R ≥ 0, will be represented by points there.

On the other hand, the complex reflection coefficient plane may be written in its polar form as:

$$\Gamma = |\Gamma|e^{\angle\Gamma} = \Gamma_r + j\Gamma_i \qquad (S1.0)$$

The magnitude of the reflection coefficient always lies between 0 and 1.0. Its angle is measured with respect to the positive real axis (Γ_r).

Theory:

The reflection coefficient is defined by:

$$\Gamma = [ZL - Zo]/[ZL+Zo] \qquad (S2.0)$$

Here ZL is the load impedance and Zo is the characteristic impedance of the source or the transmission line.
If we rearrange equation (S2.0),

$$ZL = Zo[1+\Gamma]/[1-\Gamma] \qquad (S3.0)$$

If we divide both sides by Zo we get the mapping stated above.

$$z = r + jx = [1+\Gamma]/[1-\Gamma] \qquad (S4.0)$$

Once we substitute the complex expression for the reflection coefficient and equate the real and imaginary parts we get two equations that represent circles in the complex reflection coefficient plane shown below.

$$[\Gamma_r - r/(1 + r)]^2 + [\Gamma_i - 0]^2 = [1/(1+r)]^2 \qquad (S5.0)$$

and,

$$[\Gamma_r - 1]^2 + [\Gamma_i - 1/x]^2 = [1/x]^2 \qquad (S6.0)$$

The first circle is centered at :

$$[r/(1+r), 0] \qquad (S7.0)$$

and its location is *always inside* the unit circle of the complex reflection coefficient plane. The radius of this circle is:

$$1/(1+r) \qquad (S8.0)$$

This circle is always *fully contained* in the unit circle. The radius cannot be greater than unity.

The second circle is centered at:

$$[1, 1/x] \qquad (S9.0)$$

and its location is *always outside* the unit circle in the complex reflection coefficient plane.

The centers of both circles will be to the right of the unit circle.

The radius of the second circle is $|1/x|$. This radius can vary from 0 to infinity.

The first circles centered on the real axis represent lines of *constant real part of the load impedance.* To re-iterate, r = constant and x can vary here.

The circles whose centers lie outside the unit circle represent lines of *constant imaginary part* of the load impedance. Here x = constant and r varies.

Circles centered at the match point where ZL = Zo or Γ= 0 are equidistant from the origin (i.e. $|\Gamma|$= constant). These circles are called constant VSWR circles.

Of course,
VSWR = Vmax/Vmin = $[1+|\Gamma|]/[1-|\Gamma|]$ (S10.0)
for reference.

In this equation, Vmax and Vmin are the maximum and minimum amplitudes of the standing waves created by source –

load mismatch. (Please see the description of VSWR and standing waves at the beginning of this book)

VSWR can vary from 1 to infinity.

The phase of the reflection coefficient is given by the angle from the right – hand side horizontal axis. The angle of the reflection coefficient can vary from – 180 degrees to + 180 Degrees. Angles above the horizontal axis are positive and those below are negative. Please use the Smith Chart program to view its image.

As is well known, once the VSWR is extracted other quantities such as Return Loss can be calculated as well. So VSWR is an important parameter in high frequency design. Please see the definitions of this and other parameters in the first part of this eCADbook.[TM].

There are some points on the Smith Chart with significant importance. These are listed in the table below and illustrated on the Smith Chart drawing.

Point	Name	Z - plane	Γ - plane	Comments
A	Reference impedance	$1 + j0$	$0.0\angle 0$	Single point
B	Ideal resistances	$r+j0$	$\rho\angle 0$ or $\rho\angle 180$	Main diagonal of the chart
C	Ideal capacitive reactances	$0 - jx$	$1.0\angle$Any -	Lower half of the circumference (No resistance)
D	Ideal inductive reactances	$0 +jx$	$1.0\angle$Any +	Upper half of the circumference (No resistance)
E	Short circuit	$0 +j0$	$1.0\angle 180$	Single point
F	Open circuit	∞	$1.0\angle 0$	Single point
G	Upper half of the chart	$r+jx$	$\rho\leq 1\angle$Any +	Inductive half of the chart

H	Lower half of the chart	r – jx	ρ≤1∠Any -	Capacitive half of the chart
I	A specific z value	r+ jx	ρ∠+Angle	Single point
J	Complex conjugate of z value	r - jx	ρ∠-Angle	Single point

TABLE 1.0 Significant points on a Smith Chart

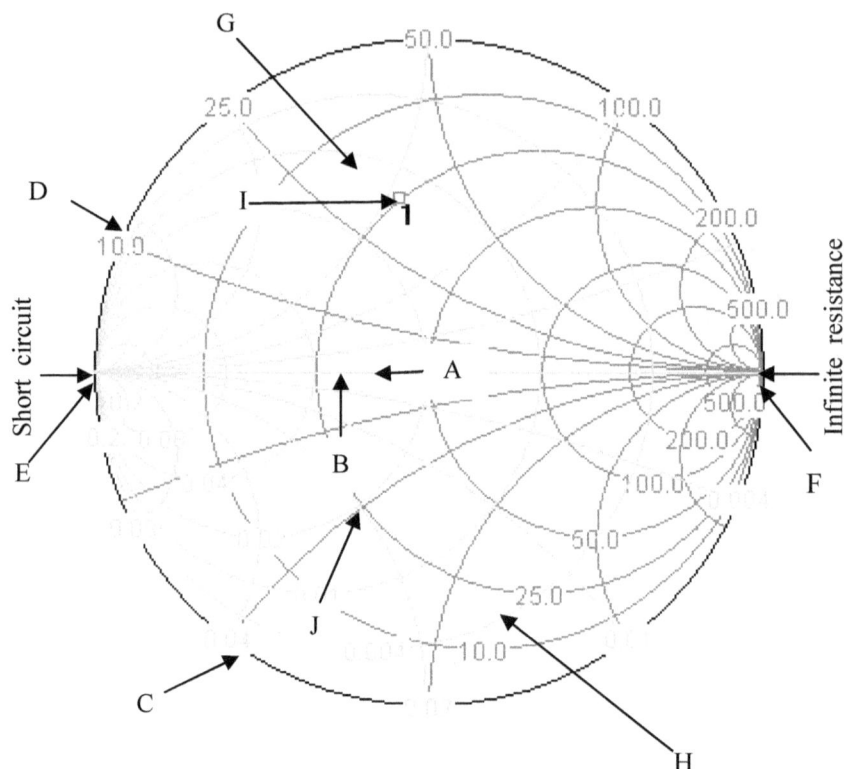

Figure S3.0 Significant points on a Smith Chart:

Impedance manipulations using the Smith Chart.

The Smith Chart can be used in impedance matching. In order to do this the Smith Chart shows, graphically, the locus of the points when adding series and shunt impedances. This section of the eCADbook^{TM} shows the techniques.

The analysis shown in this section relies heavily on a demo version Smith Chart program. The particulars of the program are:

 **V 1.91**
This program has been developed by Prof. Fritz Dellsperger, Juerg Tschirren and Roger Wetzel
© 1995 - 2000 by Berne Institute of Engineering and Architecture

It was downloaded from the web as freeware.

─Licence─────────────────────────────────
No valid licence. This copy of 'smith.exe' runs as a DEMOVERSION.

Start with a Smith Chart with a single datapoint. This is where the matching or manipulation starts from. The start is shown below as datapoint "1". This datapoint represents the impedance to be matched to the 50 Ohm (or other impedance) reference load.

Fundamentally datapoint 1, is the source impedance to be matched to the reference impedance of 50 Ohms. On the Smith Chart this 50 Ohm impedance lies at the center of the chart as explained and shown above.

The idea is to use "lossless" elements to *move the datapoint to the center of the chart.* Once this is accomplished the matching problem has been solved and the source and the load are conjugate matched to each other.

This is the effort throughout the matching techniques using the Smith Chart.

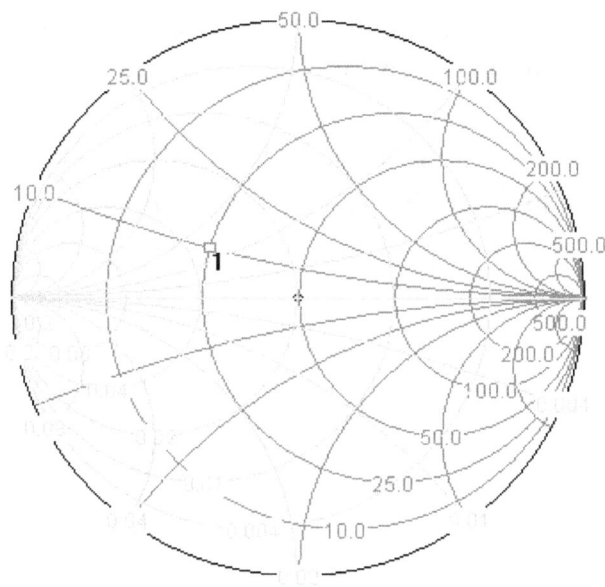

Figure S4.0. Starting from an impedance
shown as point "1" on the chart above

This datapoint has the following characteristics: (Reference
resistance is 50 Ohm)

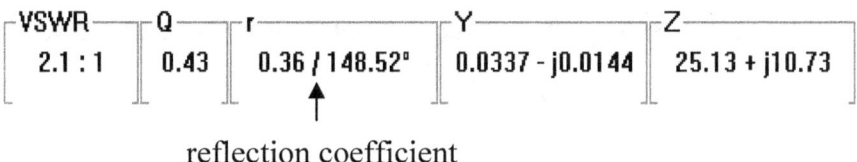

VSWR	Q	r	Y	Z
2.1 : 1	0.43	0.36 / 148.52°	0.0337 - j0.0144	25.13 + j10.73

reflection coefficient

We want to add a series inductor to this. This is shown graphically below.

We added a 2.4 nH inductor to the series circuit shown below also. Graphically our datapoint now moves to the new point "2" shown.

Note that the movement is upwards along a circle of positive reactance (i.e an inductor) and a constant resistor.

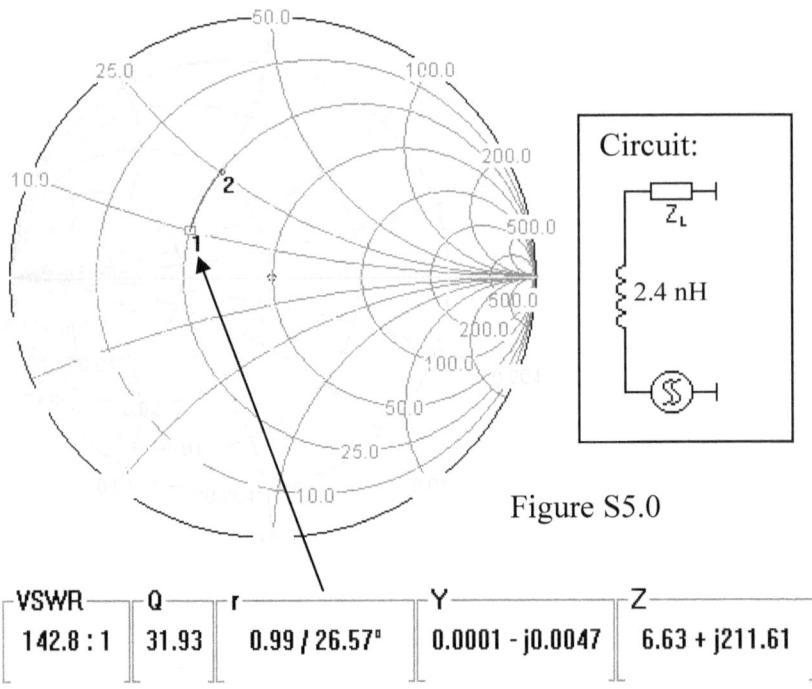

Figure S5.0

VSWR	Q	r	Y	Z
142.8 : 1	31.93	0.99 / 26.57°	0.0001 - j0.0047	6.63 + j211.61

This example shows the addition of a series inductor. If a series capacitor is added as shown below, the trajectory of the datapoint is as shown. Note, the movement is along the negative reactance circle to point "2"

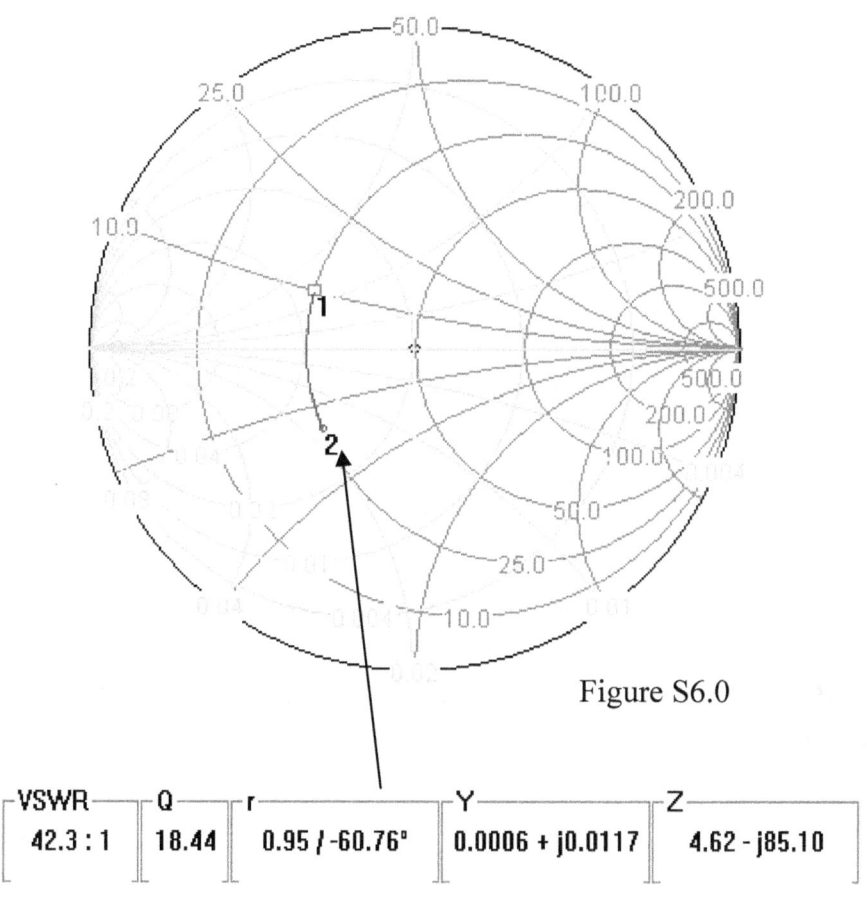

Figure S6.0

VSWR	Q	r	Y	Z
42.3 : 1	18.44	0.95 / -60.76°	0.0006 + j0.0117	4.62 - j85.10

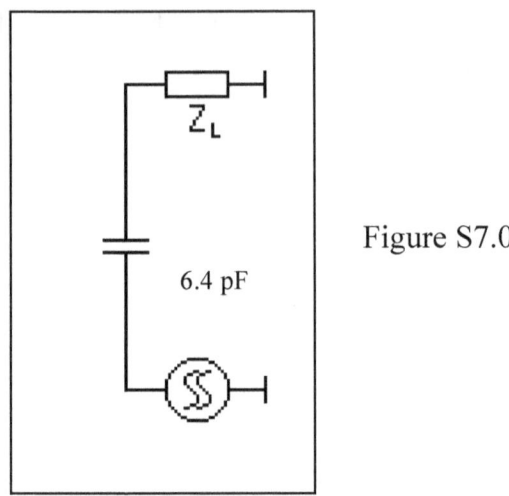

Figure S7.0

Note the sign of the reactance in the two cases. One is positive (for the addition of an inductor) and the other is negative by virtue of adding a capacitor in series. The resistance is still on a constant circle.

The match is so bad that the VSWR's are in the stratosphere!

Finally a series resistor is added. This moves the datapoint as shown below.

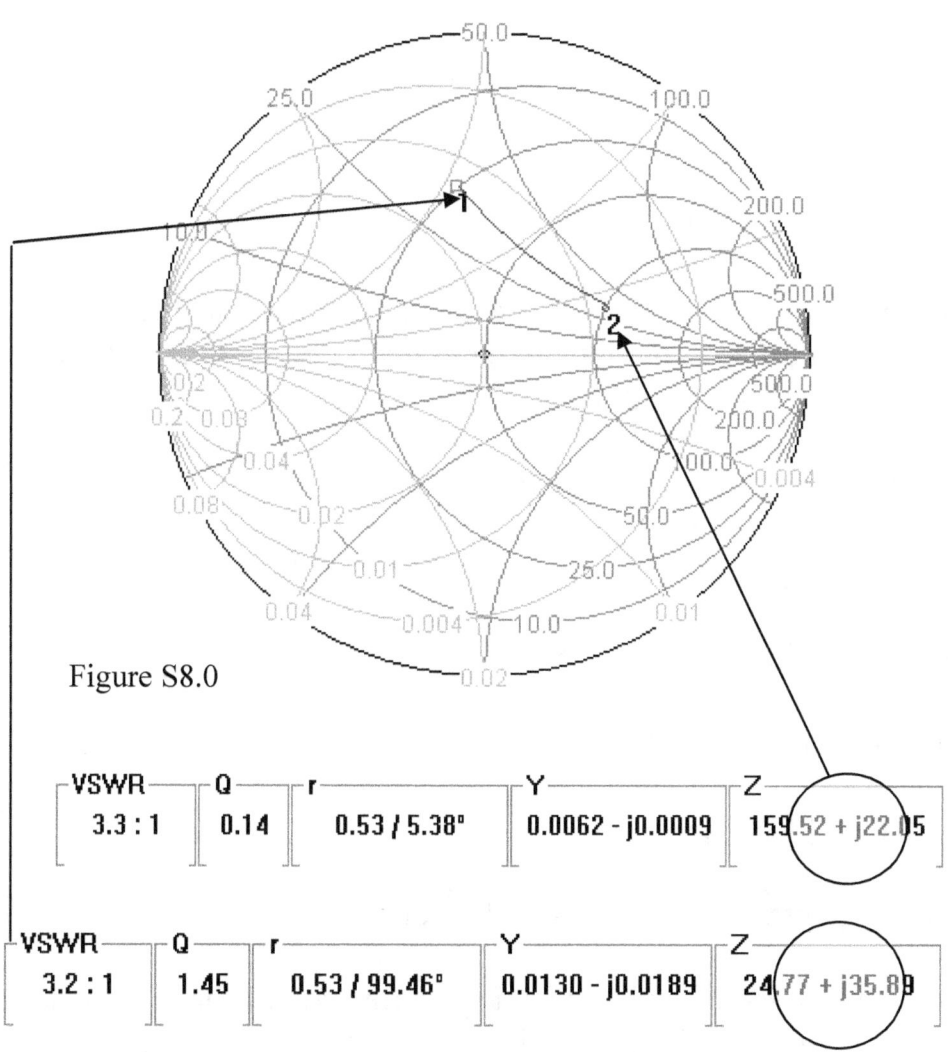

Figure S8.0

VSWR	Q	r	Y	Z
3.3 : 1	0.14	0.53 / 5.38°	0.0062 - j0.0009	159.52 + j22.05

VSWR	Q	r	Y	Z
3.2 : 1	1.45	0.53 / 99.46°	0.0130 - j0.0189	24.77 + j35.89

The datapoint has moved from r = 24.77 to r = 159.52 by adding a series resistor.

The movement is from constant resistance circle to constant resistance circle.

Admittance chart:

So far the examples shown have used the impedance chart. the Smith Chart also provides *an admittance chart*. This is shown in the figures of the Smith Chart above.

From the basic theory of normalization:

$$\Gamma = [Z - Zo]/Z + Zo] = [1/Y - 1/Yo]/[1/Y + 1/Yo] \quad (S11.0)$$

i.e replace the impedance by the admittance.

Then normalize to get,

$$\Gamma = [1 - y]/[1 + y] \qquad\qquad (S12.0)$$

The normalizing admittance is 1/Rref.

The admittance chart is a 180 degree rotation of the impedance chart by virtue of the expression for impedances above. Here we have 1 − y/1 + y instead of [z − 1]/[z + 1].

Thus the admittance line and circles in the chart below represent the admittance chart. Note, the conductance is the reciprocal of the corresponding impedance.

Figure S9.0

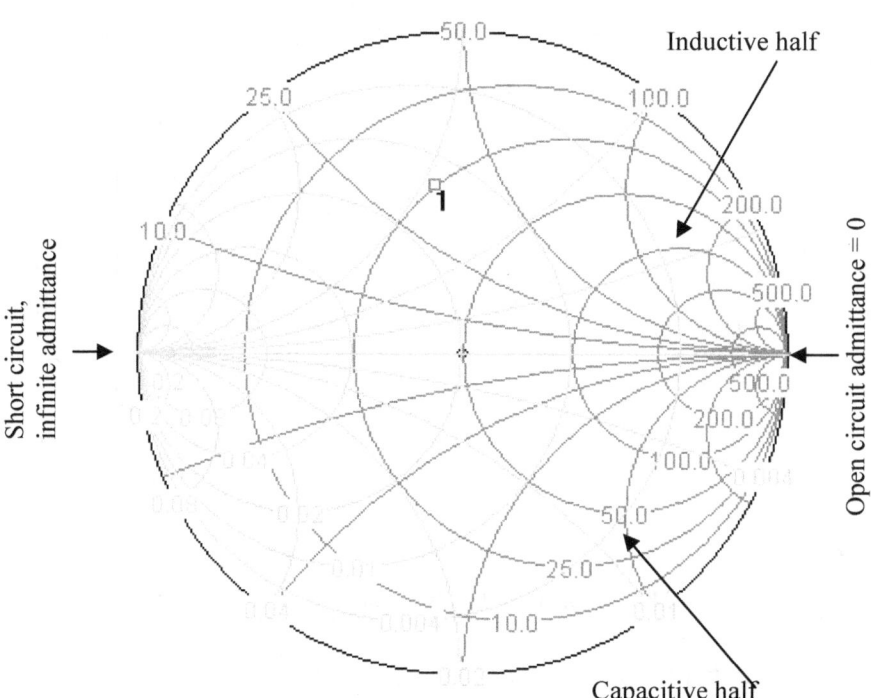

Also,

Inductive susceptance is in the top half and capacitive susceptance is in the lower half of the admittance chart. *Inductive susceptances are negative* numbers, while *capacitive susceptances are positive* numbers.

The following set of figures shows the effect of adding shunt elements.

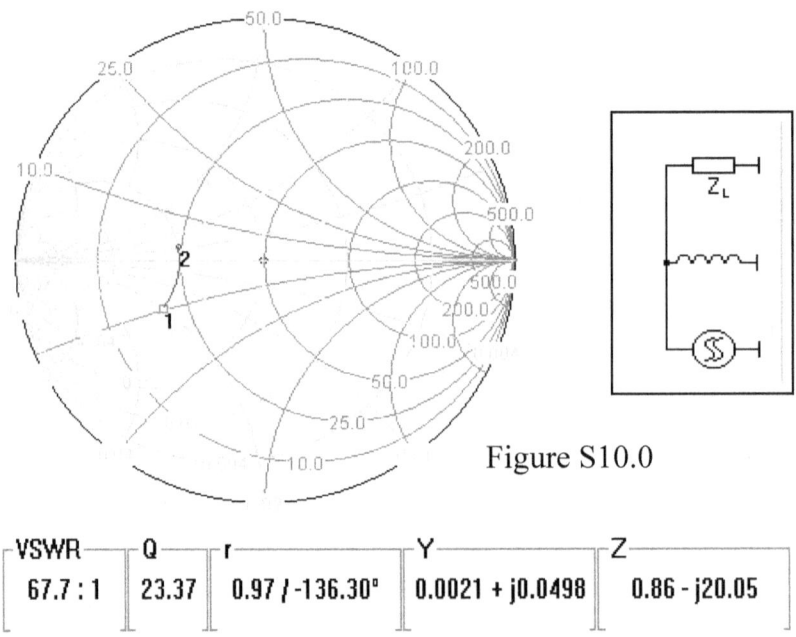

Figure S10.0

VSWR	Q	r	Y	Z
67.7 : 1	23.37	0.97 / -136.30°	0.0021 + j0.0498	0.86 - j20.05

Added shunt inductance. Movement is from "1" to "2". Reference Yo is 0.02

Point "2" parametrics.

VSWR	Q	r	Y	Z
2.6 : 1	0.50	0.45 / -153.79°	0.0404 + j0.0201	19.84 - j9.89

Point "1" parametrics

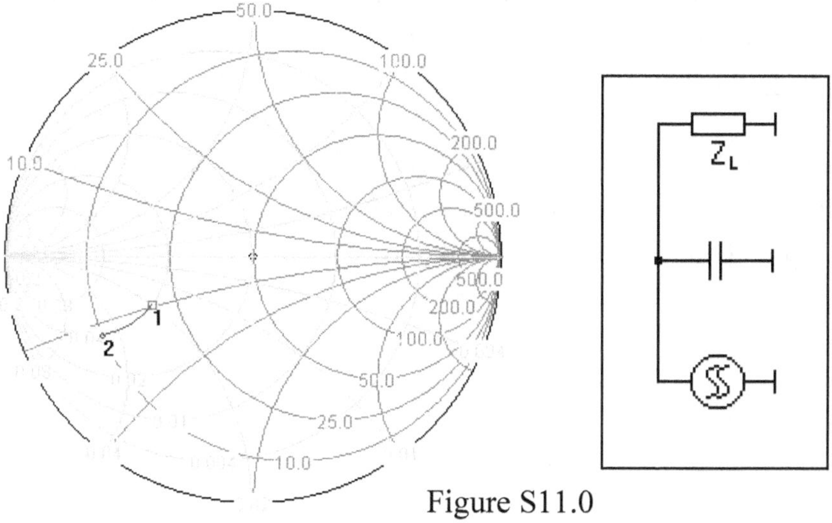

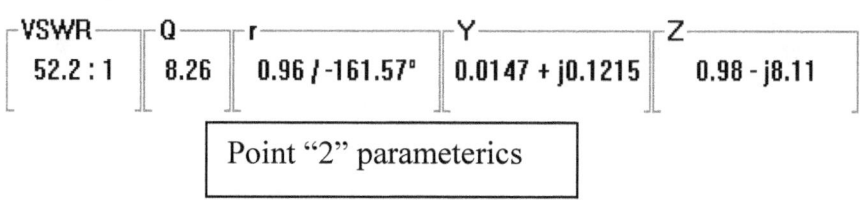

Figure S11.0

Added shunt capacitance. Movement is again from "1"to "2"

VSWR	Q	r	Y	Z
52.2 : 1	8.26	0.96 / -161.57°	0.0147 + j0.1215	0.98 - j8.11

Point "2" parameterics

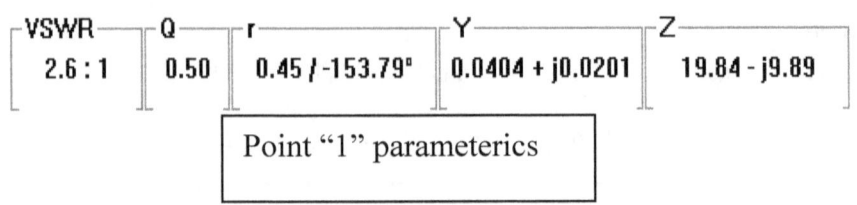

Point "1" parameterics

From the above graphics construction it is clear that *shunt inductors and capacitors move the datapoints on constant conductance circles*

The next graph shows the effect of adding a shunt resistor. Increasing the value of a shunt resistor results in movement of the datapoint towards infinite admittance, i.e a short circuit.

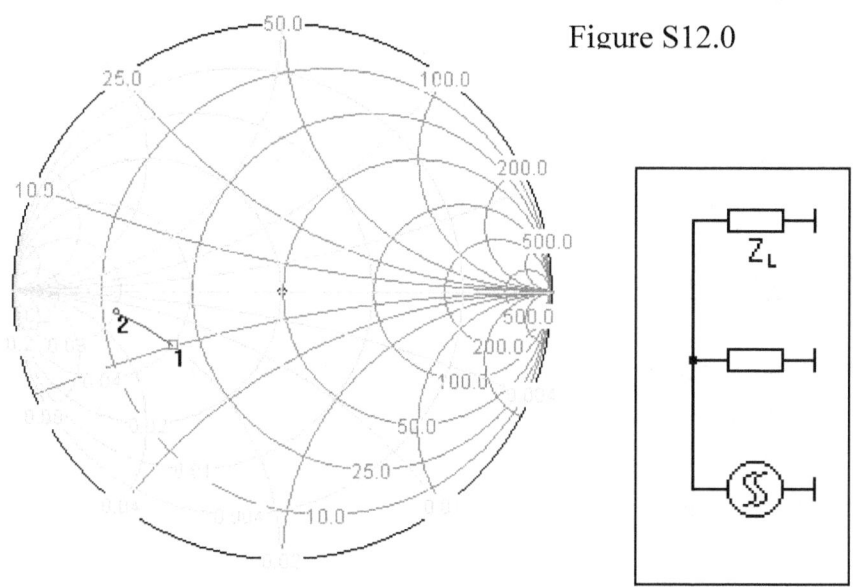

Figure S12.0

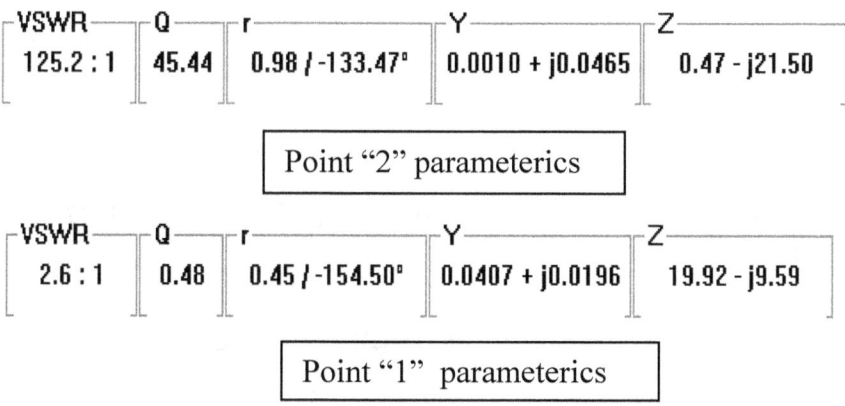

Added a shunt resistor. The datapoint moves from "1" to "2" which is in the direction of infinite admittance.

VSWR	Q	r	Y	Z
125.2 : 1	45.44	0.98 / -133.47°	0.0010 + j0.0465	0.47 - j21.50

Point "2" parameterics

VSWR	Q	r	Y	Z
2.6 : 1	0.48	0.45 / -154.50°	0.0407 + j0.0196	19.92 - j9.59

Point "1" parameterics

Transmission line manipulations:

One of the interesting features of the Smith Chart, is the manipulations that can be done on transmission lines. Again the following figures show effects of manipulating transmission lines on the Smith Chart.

Cascade transmission line

A few words of explanation are in order at this point when we start examining transmission line manipulations on the Smith Chart. The simplest construct is a *cascade transmission line* shown in the figure below.

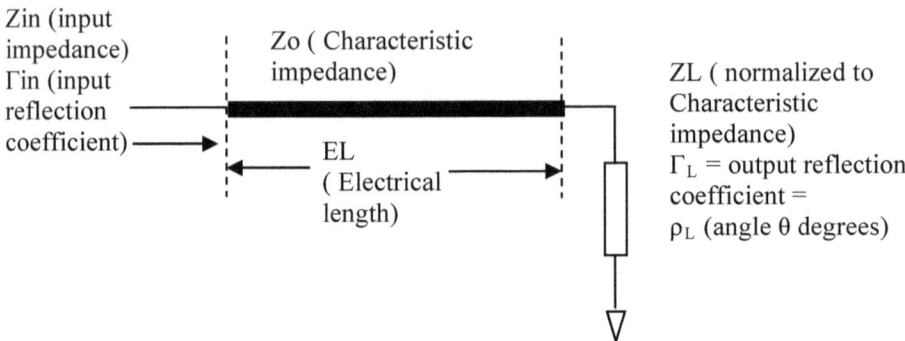

The transmission line has an input impedance given by the familiar equation:

$$ZIN = Zo[ZL+jZo\tan(\theta)]/[Zo+jZL\tan(\theta)] \qquad (S13.0)$$

Normalizing this by Zo gives,

$$zin = [zL + j\tan(\theta)]/[1+jzL\tan(\theta)] \qquad (S14.0)$$

The reflection coefficient is given by (normalized form):

$$\Gamma = [z - 1]/[z+1] \qquad (S15.0)$$

Substituting equation S14.0 into this expression gives, after some manipulation.

$$\Gamma_{IN} = \Gamma_L e^{-j2\theta} \qquad (S16.0)$$

or,

$$\Gamma_{IN} = \rho_L \angle(\Phi- 2\theta) \qquad (S17.0)$$

Here θ is the electrical length of the line.

The reflection coefficient at the input has the same magnitude as the output reflection coefficient but the phase angle is rotated by Φ in a clockwise direction through 2X the electrical length of the line. This expression applies to any transmission line impedance since the load termination is normalized to the characteristic impedance of the line.

Using the Smith Chart it is possible to find the input reflection coefficient of a terminated line. First we need to normalize the termination with the reference impedance, then mark the load as a datapoint on a normalized Smith Chart. Rotating from the normalized load through an angle that is equal to *2X the electrical length* of the line moves the datapoint to the input reflection coefficient.

The reason the transmission line length is doubled for this method is that a wave applied to the input goes to the output and then comes back to the input again. A distance of 2X the electrical length.

An example is required to see this much more clearly. This is shown below.

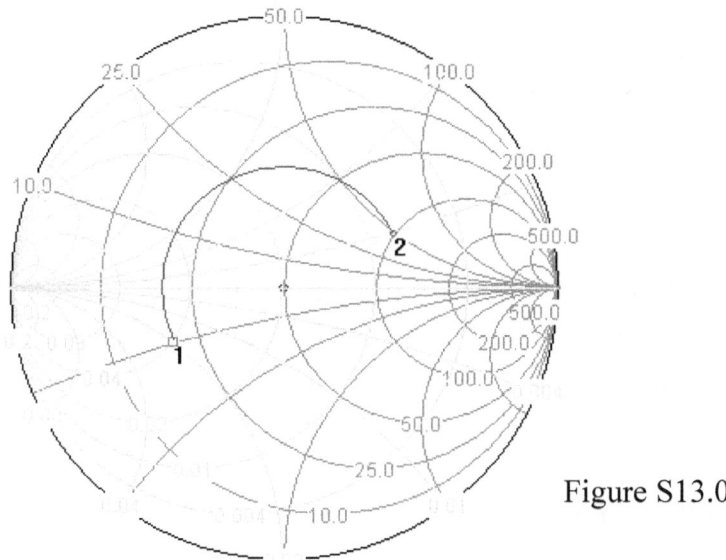

Figure S13.0

Point 2 is ¼ λ away from point 1. This means that the line length is 1/8λ. Point 2 represents the input reflection coefficient of the line. The loading is 50 Ohm and the characteristic impedance is also 50 Ohm. Loading could be different.

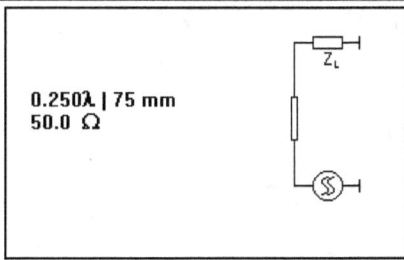

0.250λ | 75 mm
50.0 Ω

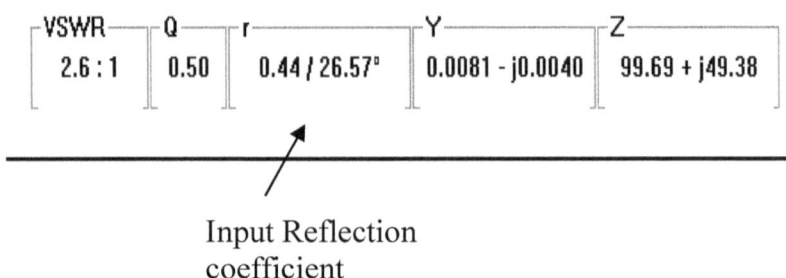

VSWR	Q	r	Y	Z
2.6 : 1	0.50	0.44 / 26.57°	0.0081 - j0.0040	99.69 + j49.38

Input Reflection
coefficient

A simple rule to remember is that if datapoint 1 is the arbitrary load terminating a transmission line, then moving away from it on the line always results in a *clockwise rotation on the Smith Chart*.

Transmission line stubs.

Lengths of transmission line can be used instead of lumped elements in a circuit for impedance matching or otherwise. Shunt transmission line stubs are an effective way of producing inductive and capacitive effects.

The two kinds of shunt stubs are *open circuited* stubs and *short circuited* stubs. These are shown schematically below.

Transmission line

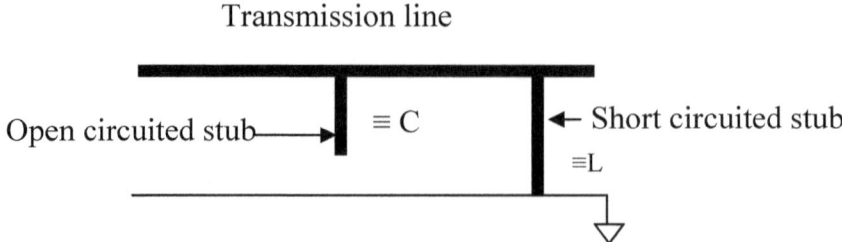

Open circuited stub ⟶ ≡ C   ← Short circuited stub
                                 ≡L

Some interesting facts about terminated and non terminated transmission lines are listed below.

TR1. Very short (with respect to the wavelength of the signal) transmission lines, terminated with a low impedance behave *like inductors.*

TR2. Very short (with respect to the wavelength of the signal) transmission line segments terminated with high impedances or open circuits behave *like capacitors.*

The following expressions can be used to understand the effects.

The input impedance of a transmission line terminated with load ZL is given by: (*Where Zin = R + jX*). Also,

$$ZIN = Zo[ZL + jZo\tan\theta]/[Zo + jZL\tan\theta] \qquad (S18.0)$$

If ZL = 0, then ZIN becomes,

$$ZIN = Zo[jZo\tan\theta]/[Zo] \qquad (S19.0)$$

or:

$$ZIN = jZo\tan\theta \qquad (S20.0)$$

This is purely reactive. Lets call the impedance of the line Zss for a short circuited segment and its electrical length θss. In this case,

Xss the reactive part of the input impedance = Zss(tanθss).

The input susceptance is simply the reciprocal of the reactance,

or:

$$Bss = 1/[Zss(tanθss)] \qquad\qquad (S21.0)$$

Now if the termination is an open circuit we get:

$$Xos = -jZos/tanθos \qquad\qquad (S22.0)$$

or, the susceptance is,

$$Bos = tanθos/Zos \qquad\qquad (S23.0)$$

The conclusion is that when the electrical lengths of shorted and open stubs are less than 90 degrees, they behave like shunt inductors and capacitors. The sign of tanθ changes repetitively through every 90 degrees. This implies that both open and shorted stubs can look like inductors or capacitors depending on their electrical lengths.

Transmission line facts: (Repeated for convenience)

Transmission lines that have electrical lengths *less than 90 Degrees* behave *inductively for short circuited loads* and *capacitively for open circuited loads.*

A RF short circuit can be produced *at any point* in a circuit by using a *short circuited transmission line with a half wavelength electrical length.* This short repeats itself every multiple of a half wavelengh.

A transmission line's input impedance is always equal to the termination at adjacent ends of the line if the characteristic impedance of the line is *the same as* the termination.

Transmission lines have *large transformation capabilities.* A short circuit can be transformed to an open circuit by using a *90 Degree* long line. The same is true of an open circuit. An open circuit can be transformed into a short circuit by a *90 Degree* line.

Cascaded transmission lines form *concentric circles* on a normalized Smith Chart if the load impedance is normalized to the characteristic impedance of the transmission line.

Parallel open and short circuited stubs behave *inductively or capacitively* as long as their electrical length is *less than 90 Degrees.*

Parallel stubs *always* move on the constant conductance circles on the Smith Chart.

A correctly selected combination of a *parallel stub and cascade transmission line* can be used to *transform any point* on the Smith Chart to any other point. The topology of this combination is dependent on the relationship of the two points. Sometimes the cascade line can be used first followed by the stub, while at other times the stub is followed by the cascaded line.

The Immitance Chart:

It is possible to use both series elements and shunt elements on the Smith Chart. In order to do so, one needs either, (1) an impedance chart for the series elements and (2) a separate admittance chart for the shunt elements.

What is actually done is that the impedance and the admittance charts are combined. This combinational chart is called an *Immitance Chart.* Generally immitance charts come in two colors to reduce the possibility of error. The impedance chart is drawn in one color while the admittance chart is drawn in a different

color. Note that the examples given above are also on the
immitance chart. (Except the color is simply black and white)

On immitance charts, impedance movements of ideal lumped
elements always occur on the constant reactance or constant
resistance circles that intersect the starting impedance.
Admittance movement always occurs on constant susceptance or
constant conductance circles that intersect the starting point.

Once inside the immitance chart, only *negative resistance* or
negative conductance can take us *outside* the chart. There may
be values that can take us to the circumference of the chart, but
only negative values of resistance or conductance can take us
beyond that limit.

On the circumference of the chart, the reactance is always equal
to the reciprocal of the susceptance. On the main diagonal of the
chart the conductance is always equal to the reciprocal of the
resistance.

In series circuits, when the circuits have lossless elements such as
inductors and capacitors, the immitance chart plots fall on the
circumference where the resistance is zero. When the circuit has
losses, the magnitude of the reflection coefficient is less than
unity and the movement occurs on constant resistance or constant
conductance circles.

Similarly in the case of parallel circuits, movement is on the constant conductance circles for lossy circuits and on the circumference for lossless circuits.

In order to more fully understand operations using the Smith Chart the following examples are presented. Note that these examples are presented using a Smith Chart program from the web. The idea is to present the technique. There is no need of the software and these manipulations can be done by hand. However, the question is – why? Therefore the approach that we have adopted here is to use software to do the more mundane tasks like normalization and drawing.

Example 1.

Find ZL when the reflection coefficient is found to be 0.5 (from a Vector Network Analyzer for example). The reference impedance is 50 Ohms. The frequency is 500 Mhz.
Obviously one can use the equation for the reflection coefficient,

$$\Gamma = [ZL - Z0]/[ZL+Zo]$$

or software to do this. Using the Smith Chart software given provides us with the following figures and solutions.

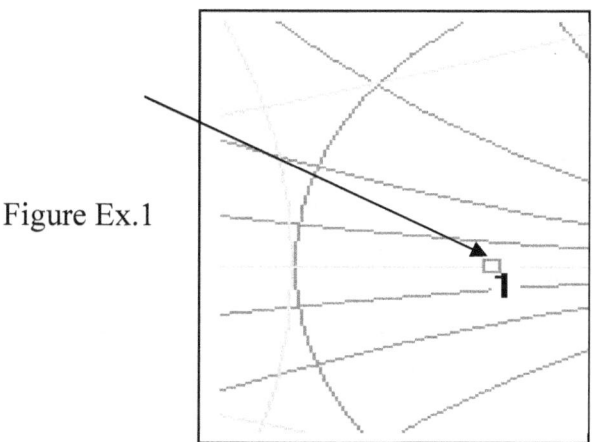

Figure Ex.1

The solution of the problem in example 1.0 was generated by using the Smith Chart software, inserting a datapoint at the value of the reflection coefficient specified and reading the impedance from coordinates generated by the software. Thus when the reflection coefficient is 0.5 with an angle of 0.0 degrees the impedance is 150.0 Ohms as can be seen above.

Further results indicate a VSWR of 3.0 and an admittance of $0.0067 - j0$.

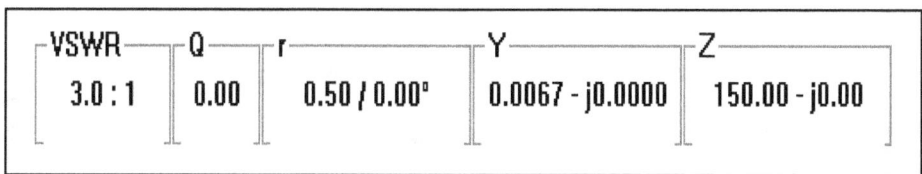

VSWR	Q	r	Y	Z
3.0 : 1	0.00	0.50 / 0.00°	0.0067 - j0.0000	150.00 - j0.00

Example 2:

Find the load impedance when the reflection coefficient is -0.3 + j0.4. Here we first convert the reflection coefficient value to polar coordinates as shown below:

The polar coordinates are shown in the results obtained from a *converter script.* Using these coordinates and the Smith Chart program we get following results.

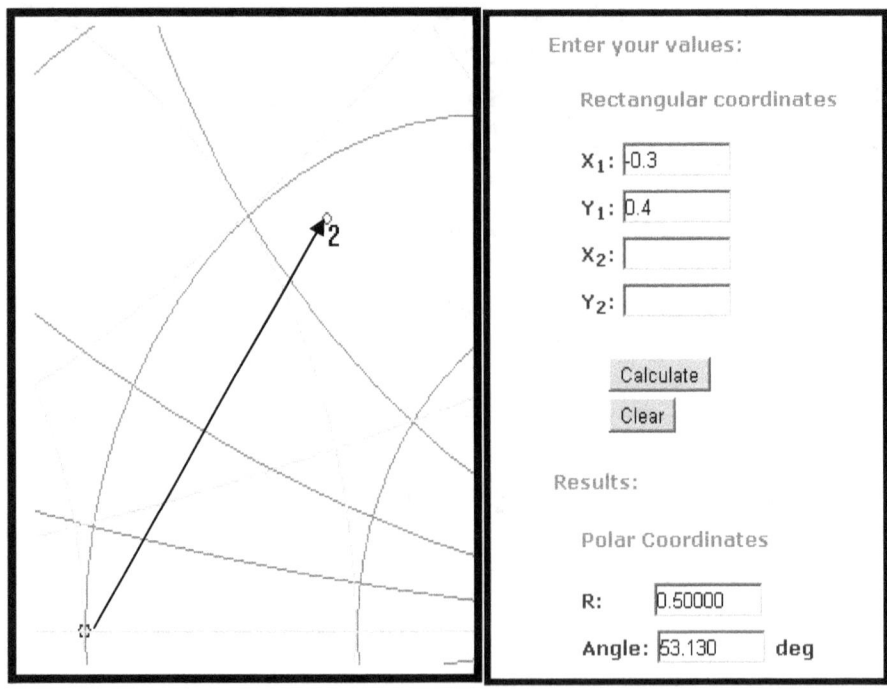

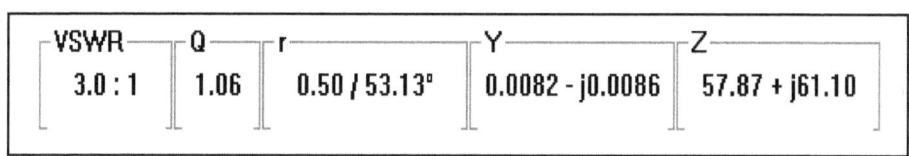

VSWR	Q	r	Y	Z
3.0 : 1	1.06	0.50 / 53.13°	0.0082 - j0.0086	57.87 + j61.10

Figure Ex 2.0

Example 3:

Match 100 Ohms to 50 Ohms at 500 Mhz.

A 50 Ohm source needs to be matched to a 100 resistor. The simplest way to do this would be to put 100 Ohms in shunt. However, this would mean unnecessarily dissipating power. Therefore we use "lossless" inductors and capacitors to do the matching.

Point 1 is at 100 Ohms.
Point 3 is at the match
point i.e. 50 Ohms.
In this matching approach,
first a capacitor was added,
(point 2)then an inductor.
The circuit is shown below.
The first move was on the
admittance chart. The second
move was on the impedance
characteristic.

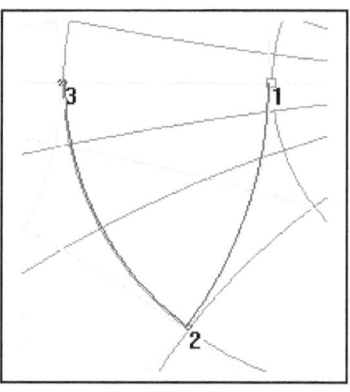

Figure Ex 3.0

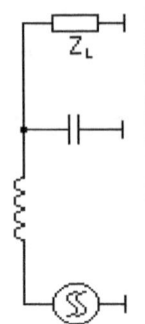

The capacitor is 3.2pF and the inductor is 16 nH.

It should be obvious that another approach will also work. In this case however, the circuit becomes high pass and DC information cannot be passed between the load and source. The circuit is shown below.

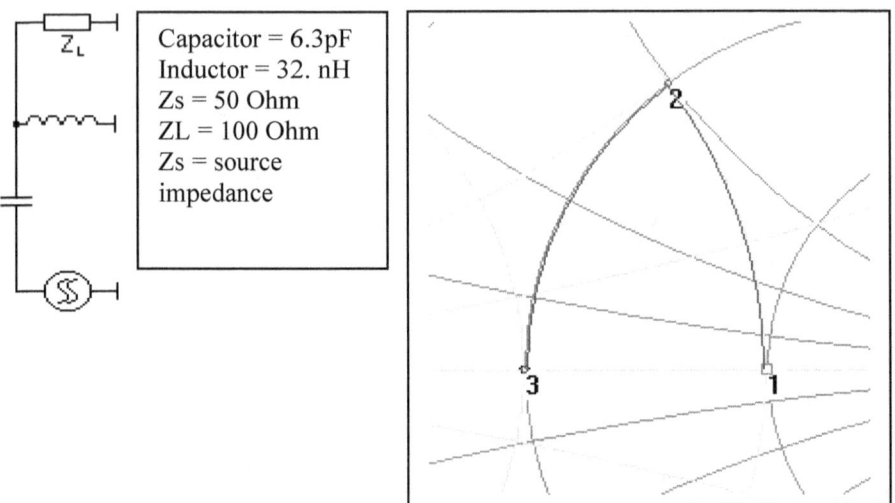

Capacitor = 6.3pF
Inductor = 32. nH
Zs = 50 Ohm
ZL = 100 Ohm
Zs = source
impedance

Figure Ex 3.1

Example 4.0:

Match a 24.97+j72.93 impedance to 50 Ohms
This solution is shown below using the Smith Chart. There are
also two moves here. The matching circuit is also shown below.

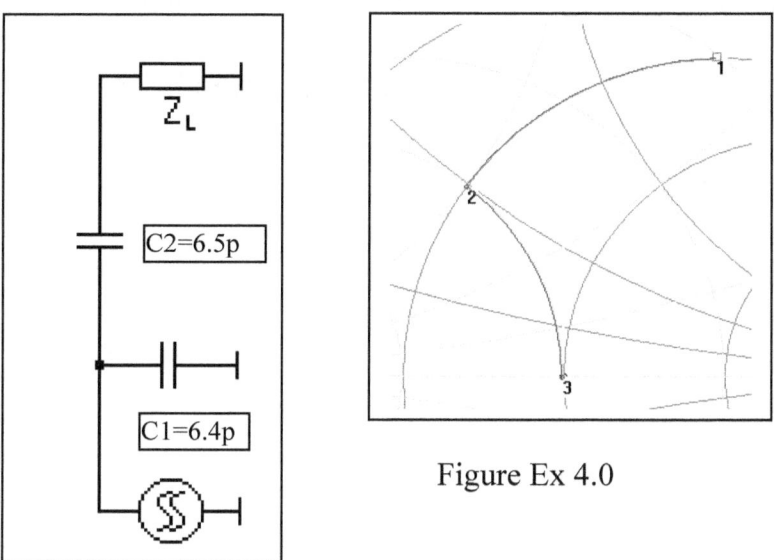

Figure Ex 4.0

These examples serve to show the process of matching using the
immitance chart. The source impedance and the load impedance
is specified. In the following examples the cascade transmission
line is used to show matching techniques.

Example 5:

This example shows how to use a combination of reactance and a cascade transmission line to provide matching. A RF power transistor operating at 1 Ghz has an equivalent input circuit of a resistor of value Rs = 2.5 Ohms in series with an inductor of Ls = 0.88 nH. This impedance needs to be matched to 50 Ohm. We will find the matching circuitry required to do this job.

The reactance of the 0.88 nH inductor is: 5.5 Ohms. Using these values the following matching manipulation was constructed on the immitance chart.

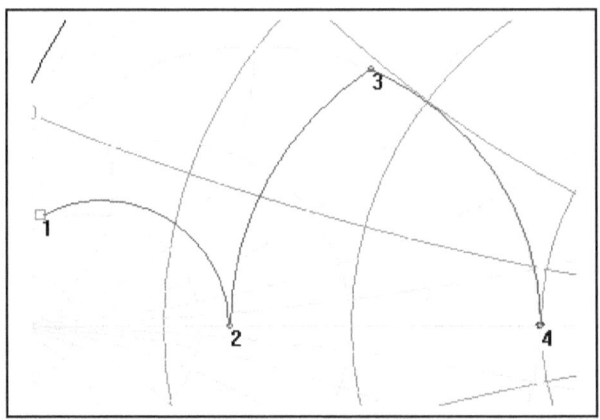

Enlarged view of the movements

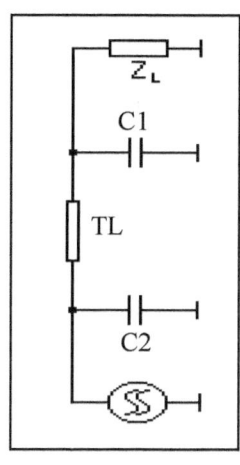

C1 = 4.1 pF
C2 = 23.8 pF
TL = 50 Ohm, 0.078λ, 23mm
ZL = 2.5 Ohms, 0.88nH

Matching circuit

Figure Ex 5.0

Example 6:

In this example we explore the use of the Smith Chart to transform an impedance ZL. We will use two lossless transmission lines to do this. The characteristic impedance is 75 Ohm for both lines. The impedance of one port is $30 - j15$ Ohms (i.e. the load).

One solution to this is shown below.

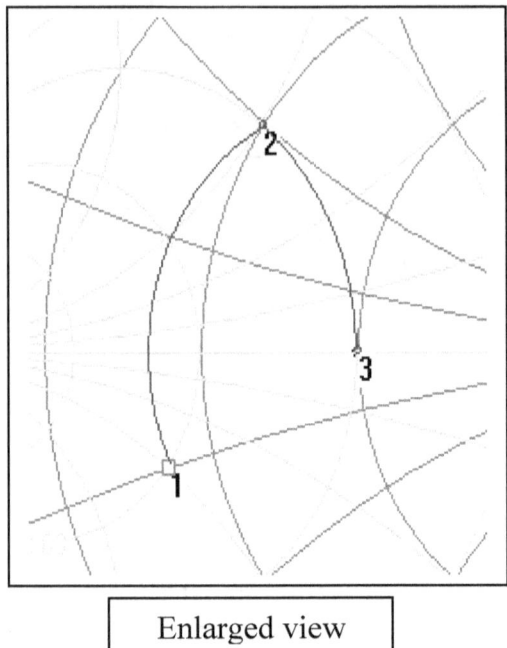

Enlarged view

Figure Ex 6.0

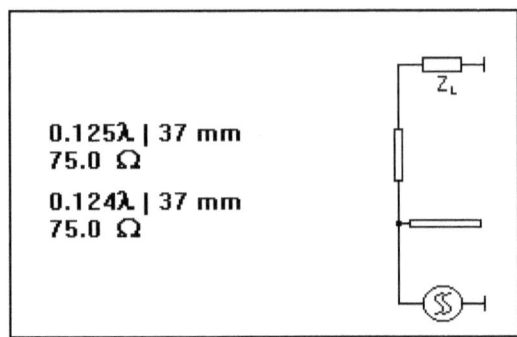

0.125λ | 37 mm
75.0 Ω

0.124λ | 37 mm
75.0 Ω

Figure Ex 6.1

As can be seen, the transformation was done by using a cascade line and a open-circuited parallel stub with an electrical length of less than 90 Degrees.

It should be obvious that there are other configurations to do this. An example is presented below.

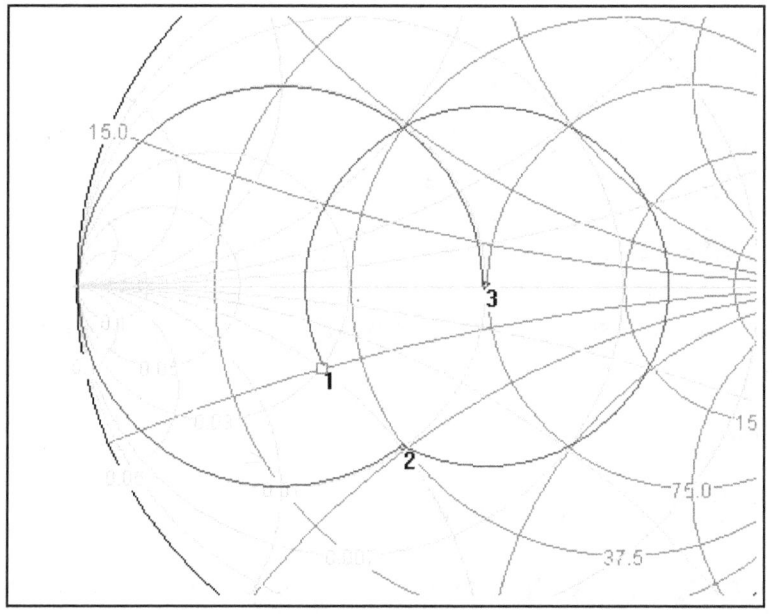

Figure 6.2

the only issue here is the length of the lines! The first configuration has shorter lines

Part VI
Appendices and Index

<u>dBm</u>: Power ratio in dBs with the reference power level of 1 mW.

$$Po(dBm) = 10Log(Pin/1e-3)$$

Po = Power in dBm
Pin= Power in Watts

Voltage can be converted to dBm. For these conversions the reference resistance is required, as shown below.

$$P(dBm) = 10log(v^2/[\ R*1e-3])$$

R is the reference resistance or load resistance, typically 50 Ohms in RF systems.

v is the rms voltage.

$$Vrms = 0.3535* V(peak\ to\ peak)$$

There are many dBm to Vrms and vice – versa converters on the World Wide Web.

Details of the Smith Chart Software used in this eCADbook.

 V 1.91
This program has been developed by Prof. Fritz Dellsperger,
Juerg Tschirren and Roger Wetzel
© 1995 - 2000 by Berne Institute of Engineering and Architecture

─Licence────────────────────────────────────
No valid licence. This copy of 'smith.exe' runs as a DEMOVERSION.

It was downloaded from the web as freeware. Permission for its use was granted by the authors.

A number of calculators and converters were used to verify the theory presented in this book. These were all available on the World Wide Web.

- Convert rectangular to polar coordinates
- L and C reactance calculators
- Calculators for complex numbers

References:

1.0   Circuit design using personal computers. Thomas R. Cuthbert Jr.,
        Wiley – Interscience Publications, John Wiley & Sons.
2.0   Fields and Waves in Communication Electronics, Simon Ramo, John
        R. Whinnery and Theodore Van Duzer. John Wiley & Sons, Second
        Edition.
3.0   Practical RF Circuit Design for Modern Wireless Systems. Volume
        I, Les Besser and Rowan Gilmore, Artech House.
4.0   Foundations of Interconnect and microtrip design. T.C. Edwards,
        M.B. Steer, third edition, J. Wiley & Sons, Ltd.
5.0   A study of microstrip design in silicon technology using closed
        form analytical expressions. Ain Rehman, Signal Processing Group
        Inc.

Javascripts and other useful programs for calculating many of the quantities described in the book are available from the publisher. Please download the scripts from http://www.signalpro.biz/impmatch.html

Any comments or error reporting should be directed to the author at spg@signalpro.biz.

www.ingramcontent.com/pod-product-compliance
Lightning Source LLC
Chambersburg PA
CBHW051324170526
45166CB00002B/672